Oxidation (IN TWO VOLUMES)

VOLUME **2**

TECHNIQUES AND
APPLICATIONS IN ORGANIC SYNTHESIS

EDITOR: *Robert L. Augustine*

CATALYTIC HYDROGENATION
by Robert L. Augustine

REDUCTION
Robert L. Augustine, EDITOR

OXIDATION, VOLUME 1
Robert L. Augustine, EDITOR

OXIDATION, VOLUME 2
Robert L. Augustine and D. J. Trecker, EDITORS

OXIDATION *(in two volumes)*

edited by

Robert L. Augustine

DEPARTMENT OF CHEMISTRY
SETON HALL UNIVERSITY
SOUTH ORANGE, NEW JERSEY

and

David J. Trecker

UNION CARBIDE CORPORATION
RESEARCH AND DEVELOPMENT DEPARTMENT
CHEMICALS AND PLASTICS
SOUTH CHARLESTON, WEST VIRGINIA

VOLUME **2**

1971

MARCEL DEKKER, INC., New York

MARCEL DEKKER, INC.
95 Madison Avenue, New York 10016

LIBRARY OF CONGRESS CATALOG CARD NUMBER: 68–12550

ISBN NO.: 0–8247–1023–1

PRINTED IN THE UNITED STATES OF AMERICA

Introduction to the Series

The synthetic organic chemist, must, of necessity, be well versed in the applications and subtleties of a wide variety of reactions. As time goes on and the volume of literature expands, it becomes increasingly difficult for the practicing organic chemist to be aware of all of the applications of a given reaction. It also becomes much more difficult for him to select the conditions which are most suitable for each particular application of a given reaction. It is the purpose, therefore, of this series on techniques and applications in organic synthesis to provide chemists with concise and critical evaluations of as many reactions of synthetic importance as possible.

Each volume deals with a single operation (hydrogenation, reduction, etc.), so that the information is readily available. Each chapter has been written by a chemist familiar with the various ramifications of the subject. Specific recommendations as to the most general or applicable experimental procedures are presented in bold-face type. Mechanistic considerations are included in the discussions primarily in order to enhance the understanding of the particular reactions. It was not intended for the chapters to serve as comprehensive reviews of the literature, so that only those sources which were either particularly representative or which contained unique or interesting aspects are included in the reference sections.

A number of topics which could have been included have been omitted from each volume. In many instances topics were not included either because the reaction in question was not in general use or because a recent review giving the synthetic applications of the reaction was available. In other instances, particularly concerned with reactants, coverage was omitted because a realistic evaluation of the extent of their general synthetic utilization was not available. It is planned, however, to have such reactions included in later volumes if their synthetic usefulness so warrants.

It is hoped that the series will provide the synthetic organic chemist with a valuable tool for carrying out complex syntheses, and that through its use he will have more time to devote to the attainment of his final goal.

> "New parts must then be added: 'Twill follow next
> If thou percase would'st vary still its shapes,
> That by like logic each arrangement still
> Requires its increment of other parts."
>
> Lucretius *Rerum Novarum*

South Orange, New Jersey
August, 1967

ROBERT L. AUGUSTINE

Preface

The contents of this second volume in the series dealing with oxidation —its technique, scope, and limitations—were chosen largely on the basis of topical interest. All chapters deal with "reactions of the sixties"—oxidation genres which gained particular significance during the last decade.

- Sulfoxide–carbodiimide reactions—a selective oxidation mode presented in experimental review by one of its principal discoverers.
- Photooxygenation—an old technique which became a finely honed synthetic tool in the sixties.
- Hydroperoxide-based epoxidations—a unique process of considerable recent importance in industry.
- Metal ion–peroxide oxidations—a broad reaction class reviewed with an eye to its particular features of synthetic utility.

As in earlier volumes in the series, each chapter has been written by an experienced practitioner in the field. Literature reviews are included primarily as a means of defining the scope and useful limits of each oxidation class. Likewise, discussions concerning reaction mechanism are intended as adjuncts to explain preferred experimental conditions and to provide guidelines for the "how" of conducting the reaction.

With the subtleties of organic synthesis multiplied almost daily by the outflow of chemical literature, the synthetic chemist must increasingly turn to concise, single references on contemporary techniques. It is hoped that this volume will fill such a need.

South Orange, New Jersey ROBERT L. AUGUSTINE

South Charleston, West Virginia DAVID J. TRECKER

July, 1970

Contributors to Volume 2

W. R. ADAMS, Union Carbide Corporation, Chemicals and Plastics, South Charleston, West Virginia

A. R. DOUMAUX, JR., Union Carbide Corporation, Chemicals and Plastics, South Charleston, West Virginia

RICHARD HIATT, Department of Chemistry, Brock University, St. Catharines, Ontario, Canada.

J. G. MOFFATT, Institute of Molecular Biology, Syntex Research, Palo Alto, California

Contents of Volume 2

INTRODUCTION TO THE SERIES iii

CONTRIBUTORS TO VOLUME 2 v

Chapter 1 **SULFOXIDE–CARBODIIMIDE AND RELATED OXIDATIONS** **1**
J. G. Moffatt

I. Introduction 1
II. Sulfoxide–Carbodiimide Oxidation of Alcohols 5
III. Dimethyl Sulfoxide–Acetic Anhydride Oxidation of Alcohols 42
IV. Oxidations with DMSO and Inorganic Acid Anhydrides 54
V. Summary 58
References 58

Chapter 2 **PHOTOSENSITIZED OXYGENATIONS** **65**
W. R. Adams

I. Introduction 65
II. Methods of Generating Singlet Oxygen 69
III. General Techniques 71
IV. Photooxygenation of Conjugated Dienes 77
V. Photooxygenation of Heterocyclic Pentadiene Derivatives 81
VI. Photooxygenation of Acenes 92
VII. Photooxygenation of Olefins 94
References 109

Chapter 3 **EPOXIDATION OF OLEFINS BY HYDROPEROXIDES** **113**
Richard Hiatt

I. Introduction 113
II. Reaction Conditions 117

III. Experimental Procedures 131
IV. Kinetics and Mechanism 134
References 138

Chapter 4 **METAL ION-CATALYZED OXIDATION OF ORGANIC SUBSTRATES WITH PEROXIDES** **141**
A. R. Doumaux, Jr.

I. Introduction 141
II. Scope and Mechanism 143
III. Reaction Conditions 158
Appendix 177
References 182

AUTHOR INDEX 187
SUBJECT INDEX 199

I

Sulfoxide–Carbodiimide and Related Oxidations

J. G. MOFFATT
INSTITUTE OF MOLECULAR BIOLOGY
SYNTEX RESEARCH
PALO ALTO, CALIFORNIA

I. Introduction 1
II. Sulfoxide–Carbodiimide Oxidation of Alcohols 5
A. Determination of Optimal Reaction Conditions 5
B. Mechanistic Considerations 12
C. Scope of the Oxidation Reaction 20
III. Dimethyl Sulfoxide–Acetic Anhydride Oxidation of Alcohols . . 42
A. Introduction and Mechanism 42
B. Scope of the DMSO–Acetic Anhydride Oxidation Reaction . 44
IV. Oxidations with DMSO and Inorganic Acid Anhydrides . . . 54
A. Use of DMSO–Phosphorus Pentoxide 54
B. Use of DMSO–Sulfur Trioxide 56
V. Summary 58
References 58

I. INTRODUCTION

In recent years dimethyl sulfoxide (DMSO) has assumed a prominent position in organic chemistry due to its combination of unique solvent properties and specific chemical reactivities. The general properties of DMSO as both a solvent and a reactant have been the subject of several comprehensive reviews (*1–4*) that adequately cover the subject.

Included among the reports of DMSO reactions have been several observations of its action as an oxidizing agent (*5*). Thus Kornblum et al. (*6*) have shown that phenacyl halides react readily with DMSO at room temperature to form glyoxals or α-diketones in good yields. Less reactive halides, however, require more vigorous conditions: α-bromoesters (*7*) require temperatures of 70–100°, and primary (*8,9*) and secondary (*10*) alkyl iodides, or the corresponding tosylates, can be converted into aldehydes or ketones by heating at 150–170° in the presence of an acid acceptor such as sodium bicarbonate. Such oxidative reactions are limited in their synthetic value when applied to secondary α-haloketones since elimination frequently becomes a serious side reaction (*11,12*).

The mechanism proposed for these reactions by Hunsberger and Tien (*7*) involves nucleophilic displacement of the halide by DMSO with formation of an alkoxysulfonium ion (**1**) which subsequently loses a proton and collapses to the carbonyl compound and dimethyl sulfide as shown in Eq. (1).

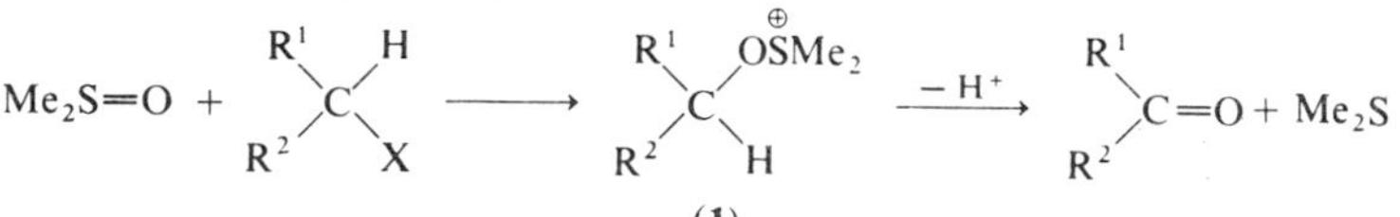

$$Me_2S{=}O + R^1R^2C(H)X \longrightarrow R^1R^2C(H)\overset{\oplus}{O}SMe_2 \ \textbf{(1)} \xrightarrow{-H^+} R^1R^2C{=}O + Me_2S \qquad (1)$$

The intervention of alkoxysulfonium ions such as (**1**) is a common feature of many of the oxidative reactions involving DMSO. For example, they have been proposed as intermediates in the oxidative reactions of DMSO with epoxides (*13,14*), chloroformates (*15*), diazonium salts (*16*), and quinol acetates (*17*).

Several years ago, during an attempt to develop a new phosphorylating agent, we wished to investigate the reaction of *N*-phosphorylpiperidine (**2**) and dicyclohexylcarbodiimide (**3**, DCC) with 2′,3′-*O*-isopropylideneuridine (**4**) (*18*). Both the *N*-phosphoryl compound and its related symmetrical pyrophosphate derivative, however, were exceedingly insoluble in pyridine, and accordingly the reaction was attempted in a mixture of pyridine and DMSO. No phosphorylation was apparent, and to rule out any adverse solvent effect, the well-known phosphorylation of (**4**) with 2-cyanoethyl phosphate and DCC in pyridine (*19*) was modified by using DMSO as a cosolvent (*20*). Once again, examination of the reaction by paper electrophoresis clearly showed that no phosphorylation had occurred when even 10% DMSO was added.

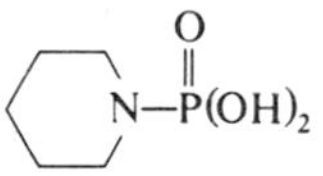

$C_5H_{10}N{-}P(O)(OH)_2$ (**2**) $C_6H_{11}N{=}C{=}NC_6H_{11}$ (**3**)

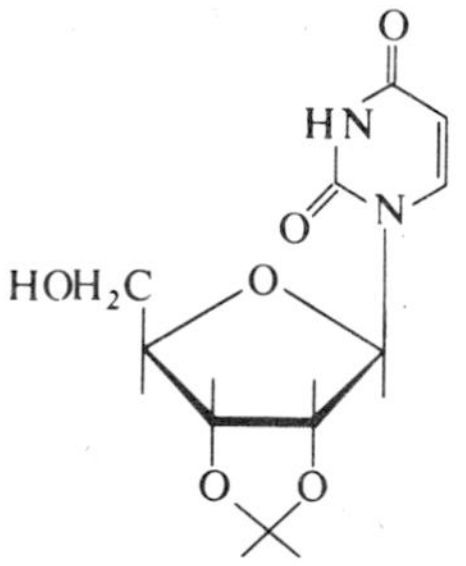

2′,3′-*O*-isopropylideneuridine, HOH_2C (**4**)

As a further test we studied the effect of added DMSO upon the well-known polymerization of thymidine 5′-phosphate (**5**) with DCC in pyridine (*21*). In the presence of even minor amounts of DMSO the reaction took a completely anomalous course; the mixture rapidly became dark colored and emitted a foul, sulfide-like smell (*20,22*). Chromatographic examination after only 15 min showed that the nucleotide (**5**) had completely disappeared, being degraded to thymine (**6**) and a mixture of inorganic polyphosphates, principally ortho- and trimetaphosphates. In addition, reactions which included pyridine gave rise to varying amounts of *N*-(thiomethoxymethyl)pyridinium ion (**7**), which was characterized by independent synthesis.

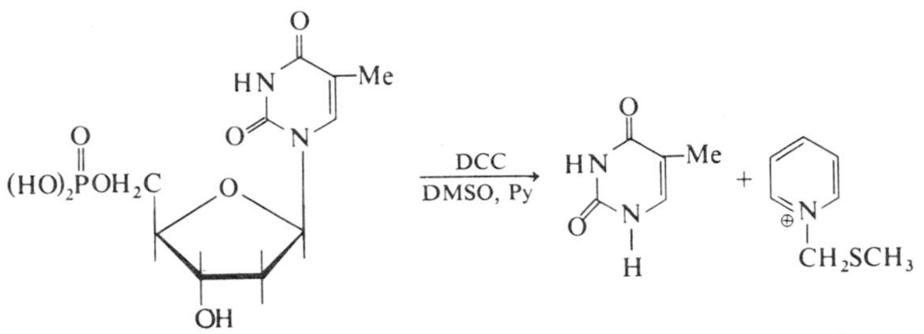

(**5**) (**6**) (**7**)

Similar degradation of other ribo- and deoxyribonucleoside 5′-phosphates via *N*-glycosidic cleavage also was observed under similar conditions (*20*). Attempts to ascertain the fate of the carbohydrate moiety fail d, with reactions upon uniformly ^{14}C-labeled nucleotides leading to a complex mixture of degradation products. To determine what structural features of the nucleotide molecule were essential for this unusual *N*-glycosidic cleavage reaction, a number of derivatives of thymidine 5′-phosphate (**5**) were examined. It soon became apparent that thymine release from (**5**) was completely suppressed if the 3′-hydroxyl function was substituted by acetyl or tetrahydropyranyl groups or if it were completely removed as in 3′-deoxythymidine 5′-phosphate. Thymidine itself was completely inert in the presence of DMSO and DCC, but upon addition of an equivalent amount of anhydrous orthophosphoric acid, an immediate reaction ensued with release of thymine. With this observation that the reaction required acidic catalysis it was then possible to examine the reactions of a number of simple, substituted nucleosides in the presence of DMSO, DCC, and anhydrous orthophosphoric acid. Such reactions showed that the release of thymine was dependent upon the presence of a free 3′-hydroxyl group, and that the presence or absence of substituents at the 5′ position was unimportant.

The most significant observation in the above experiments was that, while the reaction of 3′-*O*-acetylthymidine (**8**) with DMSO, DCC, and anhydrous orthophosphoric acid did not lead to thymine release, the reaction product was chromatographically slightly less polar than the starting material and, unlike (**8**), gave a positive test for carbonyl groups when sprayed with dinitrophenylhydrazine solution. From such reactions it was possible to isolate the crystalline dinitrophenylhydrazone of 3′-*O*-acetylthymidine 5′-aldehyde (**9**) in 61% yield.

$$\textbf{(8)} \xrightarrow[\text{DCC, } H_3PO_4]{\text{DMSO}} \textbf{(9)}$$

Later in this chapter the oxidation of nucleosides will be considered in greater detail; for the moment it is sufficient to note that the above reaction provides an extraordinarily mild method for the oxidation of an alcohol to the corresponding carbonyl compound. In particular, the oxidation of the primary hydroxyl group of a nucleoside to the aldehyde level without the formation of any detectable carboxylic acid is unique. Other attempts at oxidizing nucleosides with platinum and oxygen (*23,24*), permanganate (*25,26*), manganese dioxide (*27*), and chromic oxide (*28*) have led to the formation of 5′-carboxylates without accumulation of detectable aldehydes. Only the closely related oxidation of 2′,3′-*O*-isopropylidene-adenosine with DMSO and diphenylketen-*p*-tolimine in the presence of phosphoric acid has been reported to give the corresponding 5′-aldehyde (*29*). By analogy with the above it now appears clear that the glycosidic cleavage of nucleosides containing a free 3′-hydroxyl group is the consequence of a facile oxidation to the 3′-ketonucleoside followed by spontaneous *β*-elimination of both the heterocyclic base and the phosphate ester.

The realization that the reaction of an alcohol with DMSO, DCC, and anhydrous orthophosphoric acid leads to an efficient, yet mild, oxidation to the corresponding aldehyde or ketone has led us into a detailed study of these and related reactions. Subsequently, others have extended this type of oxidation to mechanistically closely related reactions using DMSO in conjunction with acetic anhydride (*30*), phosphorus pentoxide (*31*), and sulfur trioxide (*32*). Each of these types of reaction will be considered separately in terms of their scope and their individual advantages and limitations. An attempt has been made to cover the literature concerning these reactions as comprehensively as possible up to the spring of 1969.

Many of the combinations of reagents being discussed also can be reacted with functional groups other than alcohols leading to novel types of products (*33–36*). These aspects will not be considered in this chapter.

II. SULFOXIDE–CARBODIIMIDE OXIDATION OF ALCOHOLS

A. Determination of Optimal Reaction Conditions (*20*)

As a convenient model to quantitatively assess the effects of different reaction variables, the oxidation of testosterone (**10**) to androst-4-ene-3,17-dione (**11**) was chosen. This model offers the advantages of very facile separation of the product and reactant by thin-layer chromatography

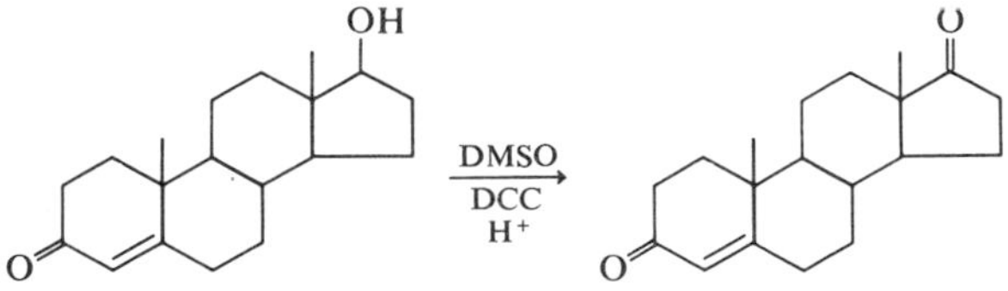

$$\textbf{(10)} \xrightarrow[\text{DCC}, \; H^+]{\text{DMSO}} \textbf{(11)}$$

and an intense ultraviolet chromophore for quantitative estimation. Except for the presence of trace amounts of two minor by-products (less than 1%) formed in certain cases, the only ultraviolet-absorbing products were (**10**) and (**11**), thus facilitating quantitative estimation of the rate and extent of the oxidation reaction.

The nature of the acid catalyst was first examined in a series of reactions between 1 molar equivalent of (**10**), 3 molar equivalents of DCC, and 0.5 molar equivalent of a variety of acids in a large excess of DMSO. The results of this study are shown in Fig. 1, from which it can be seen that a wide range of reaction rates are possible using different acids. Highly satisfactory rates of oxidation were obtained using anhydrous orthophosphoric or phosphorous acids, (**11**) being formed in close to 95% yield within the first 3 hr at room temperature. Subsequent experiments actually showed that oxidation was essentially complete within

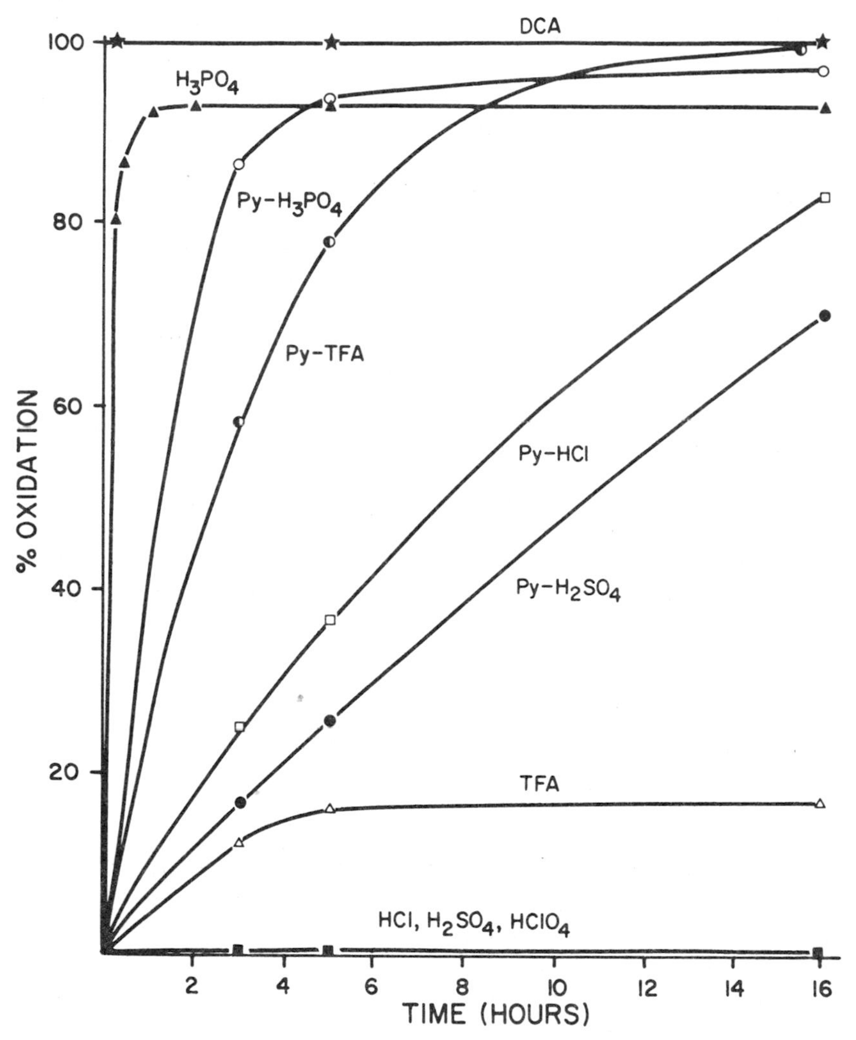

Fig. 1. Rates of oxidation of testosterone using different acids. Each reaction contained testosterone (0.1 mmole), DCC (0.3 mmole), and the appropriate acid (0.05 mmole) in DMSO (0.5 ml). The extent of reaction was measured by quantitative thin-layer chromatography of evaporated aliquots.

45 min when these acids were used. On the other hand, no oxidation was observed when mineral acids were used, and only a slow, partial reaction took place with strong organic acids such as trifluoroacetic acid. All of these strong acids, however, became satisfactory proton sources when

used as their pyridinium salts. Oxidations using the pyridinium salts of mineral acids tended to be rather slow and often incomplete, but the pyridinium salts of orthophosphoric and trifluoroacetic acids led to extremely clean reactions. For example, when pyridinium trifluoroacetate was used, a single ultraviolet-absorbing material was present following oxidation of (**10**). Trialkylammonium salts of strong acids, however, failed to provide adequate oxidation. Comparisons of the efficiencies of several other groups of acids led to some further interesting correlations. Thus, although oxidation of (**10**) to (**11**) reached a constant value of 92% in about 45 min using anhydrous orthophosphoric acid, the reaction was decidedly faster using monophenyl phosphate and reached a plateau at 90% oxidation within 15 min. On the other hand, diphenyl phosphate functioned very poorly and led to only 3% of (**11**) after 3 hr and to 37% after 5 days. A similar comparison showed that acetic acid (p*K* 4.76) and trichloroacetic acid (p*K* 0.66) did not support the oxidation of (**10**); that monochloroacetic acid (p*K* 2.86) led to slow and incomplete reaction; and that dichloroacetic acid (p*K* 1.25) gave complete oxidation with less than 1% impurities within 6 min. Our experience has been that dichloroacetic acid provides the most rapid oxidations of any acid we have tested. Indeed, it has been suggested (*37*) that in certain cases (e.g., the oxidation of estra-1,3,5(10)-triene-3,11α,17β-triol-3-methyl ether to the corresponding 11,17-dione) this acid is essential to obtain a high yield of product.

From the above it is clear that acids of intermediate strength are necessary to obtain satisfactory oxidation. Both very weak acids (e.g., acetic acid) and very strong acids (e.g., mineral acids) fail since, as will be discussed later, the reaction mechanism involves both acid- and base-catalyzed steps, and a suitable balance between the two must be reached. **Our general experience is that free orthophosphoric acid, dichloroacetic acid, and pyridinium trifluoroacetate are the most satisfactory acids for preparative purposes.** Frequently, the use of phosphoric acid leads to the formation of minor amounts (up to perhaps 5%) of by-products, usually the thiomethoxymethyl ether of the starting alcohol (e.g., (**12**) from oxidation of testosterone), whereas this side reaction is considerably reduced with pyridinium trifluoroacetate or dichloroacetic acid. On the other hand, the use of the latter two acids gives rise to a side reaction with DCC giving *N*-acylureas. Reactions using pyridinium trifluoroacetate, for example, give appreciable amounts of *N*-trifluoroacetyl *N*,*N*′-dicyclohexylurea (**13**), a very faintly ultraviolet-absorbing compound that is rather difficult to detect with many spray reagents used on thin-layer silica plates. Since acylureas such as (**13**) are very soluble in most organic solvents, their presence does not usually impede the crystallization of the desired carbonyl compounds. They are also quite nonpolar, moving with an R_f of roughly 0.5 on thin-layer silica plates using hexane–ether (1 : 1), and accordingly, are readily removed by chromatography if this is necessary. However, one should be aware of the presence of these not readily detected by-products when selecting the most appropriate acidic catalyst. In addition to the relative speeds of the oxidation reactions and the possible presence of *N*-acylureas as by-products, the acidity of the reaction mixture becomes important in certain cases. Thus, as will be seen later, the oxidation of homoallylic alcohols using anhydrous phosphoric acid is sometimes accompanied by acid-catalyzed isomerization of the double bond into conjugation with the carbonyl group produced. **This isomerization of the initial product can be largely prevented by the use of the only slightly acidic pyridinium trifluoroacetate.**

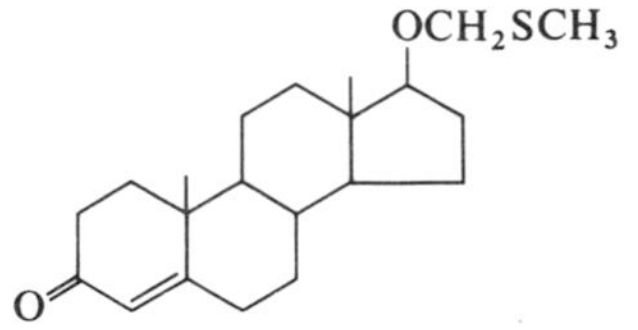

OCH_2SCH_3

(12)

$C_6H_{11}N(COCF_3)-C(=O)-NHC_6H_{11}$

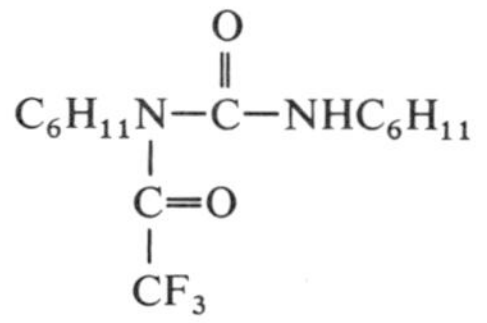

(13)

Once the nature of the acid catalyst is decided, the amount to be added is the next consideration. This question has been explored (*20*) by using, once again, the oxidation of testosterone (**10**) to (**11**) in the presence of an excess (5 molar equivalents) of DCC and from 0.1 to 2.0 molar equivalents of anhydrous orthophosphoric acid. These experiments showed that efficient oxidation accompanied the use of 0.1, 0.5, or 1.0 molar equivalent of acid, although the rate was markedly slower using only 0.1 equivalent. The use of 2 molar equivalents of acid, however, led to an initially rapid reaction which leveled off after only about 60% oxidation. This effect may well be due to depletion of the carbodiimide through side reactions (e.g., pyrophosphate formation) with the larger amount of acid. **Our experience has shown that the use of about 0.5 molar equivalent of acid provides a generally satisfactory result providing that there is not a strongly basic nitrogen function in the substrate molecule.** As mentioned above, trialkylammonium salts of various acids do not catalyze the oxidation reaction, and hence the oxidation of strongly basic molecules requires the use of greater than 1 equivalent (generally 1.5 equivalents) of acid. This point will be discussed further when dealing with the oxidation of alkaloids (Section II,C,3).

The amount of DCC required was also studied in a similar series of experiments, each containing testosterone, 0.5 equivalent of anhydrous phosphoric acid, and from 1 to 8 equivalents of DCC in excess DMSO. These experiments showed that **there was virtually no difference between the use of 3, 5, or 8 equivalents of DCC.** Use of 2 equivalents led to a markedly slower and less complete reaction, and with only 1 equivalent oxidation occurred only to an extent of 10%. A very similar effect has been noted by Jones and Wigfield (*38*) during oxidation of 3β-hydroxyandrost-5-en-17-one (**14a**) to the corresponding androst-5-en-3,17-dione (**15**). Although oxidation using 3 molar equivalents of DCC gave the desired β-γ-unsaturated ketone in 70% yield, the use of only 1 equivalent of DCC gave only 5% of (**15**) together with a similar amount of the thiomethoxymethyl ether (**14b**). The formation of the latter type of compound by a type of Pummerer rearrangement will be discussed later in the section on reaction mechanisms. The requirement for more than a molar equivalent of DCC is not mechanistically obvious but can perhaps be due in part to competing side reactions between the acid and DCC which remove the latter from reaction. The removal of excess DCC from these reactions is conveniently accomplished by addition of a slight excess of oxalic acid, which reacts rapidly with DCC to give carbon dioxide, carbon monoxide, and dicyclohexylurea (*39*). The insolubility of the latter in most organic solvents permits its efficient removal, and this method is generally satisfactory when dealing with water-insoluble compounds that are not sensitive to very mild acidic conditions.

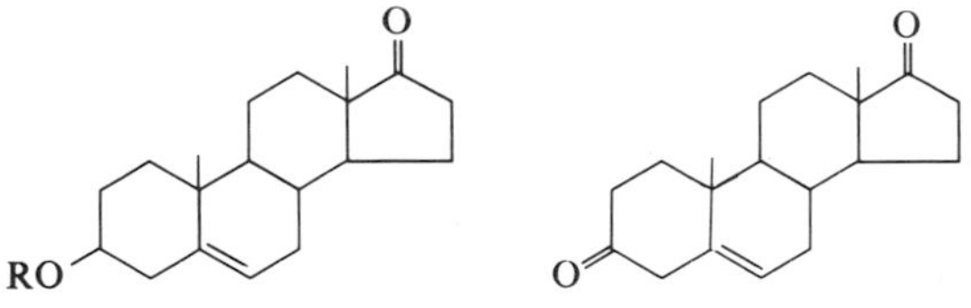

RO O O O

(**14**) (**15**)

(**a**) R = H
(**b**) R = CH_2SCH_3

The amount of DMSO necessary to effect satisfactory oxidation seems quite flexible, and essentially similar oxidations of testosterone were obtained using mixtures of DMSO and an inert solvent such as benzene containing from 10 to 100% DMSO. **For preparative oxidations we have usually employed an inert cosolvent such as benzene or ethyl acetate.** This is advantageous since DCC has a somewhat limited solubility in DMSO alone and also because it is frequently desirable to work in a not too concentrated solution. Oxidations carried out in a concentrated DMSO solution frequently become quite exothermic, and the deposition of dicyclohexylurea can give an essentially solid mixture. Also, during the work-up, the DMSO is usually removed by aqueous extraction, and in the presence of large quantities of this highly polar solvent the solubility of certain substrates in the aqueous phase becomes significant. These problems are all minimized by the use of limited amounts of DMSO admixed with an inert solvent. Other sulfoxides, such as tetramethylene sulfoxide, work equally well in the oxidation reaction, but offer no advantage, particularly since they are both more expensive and more difficult to remove by aqueous extraction following the reaction.

A variety of reagents other than DCC have been used to activate DMSO for oxidation reactions but have generally proved less satisfactory. As might be expected, other aliphatic carbodiimides behave very much like DCC. For example, diisopropylcarbodiimide is almost identical to DCC in comparable oxidation reactions, and if a choice is to be made it must be based upon the properties of the urea by-products. Thus, diispropylurea is considerably more soluble in organic solvents than is dicyclohexylurea, and, accordingly, its removal from the reaction mixture is less efficient. In some cases, however, this increased solubility offers distinct advantages during subsequent manipulations. The use of diethylcarbodiimide (*40,41*) further accentuates this difference, and diethylurea can be removed by extraction of the evaporated reaction mixture with water. In our experience, the use of diethylcarbodiimide frequently leads to incomplete oxidation reactions, and since primary aliphatic carbodiimides are unstable during prolonged storage (*40,41*) and are accordingly not commercially available, its use seems limited to special cases. **It also might be mentioned that many people show strong skin irritation upon direct contact with carbodiimides.** This should be borne in mind when using volatile carbodiimides such as the diethyl derivative and due care should be taken. Since quantitative removal of the excess carbodiimide and urea by-product is frequently a trying problem during oxidation of water-insoluble compounds, we have made several attempts to use carbodiimides containing solubilizing substituents. A number of so-called "water-soluble carbodiimides" have been developed for use in peptide synthesis. Most of these [e.g., (**16**), (**17**)] (*42,43*) contain strongly basic nitrogen functions and, as such, have proved to be rather unsatisfactory in the oxidation reaction. In an attempt to avoid the problem of basicity we have prepared a series of carbodiimides containing substituted pyridine moieties such as (**18**) (*44*). While these compounds, and the related urea derivatives, can be removed by extraction with dilute acid, they have been found to be very unstable and rather difficult to prepare. The most stable member of this series (**18**) proved to support an extremely rapid initial oxidation of several alcohols, but frequently the reaction did not reach completion even after the addition of further amounts of the various reagents. This combination of properties appears to severely limit the possible utilities of compounds such as (**18**).

$$C_6H_{11}N{=}C{=}NCH_2CH_2{-}\overset{\oplus}{N}(CH_3)(CH_2CH_2)_2O \quad MeSO_3^{\ominus}$$

(**16**)

$$C_6H_{11}N{=}C{=}N{-}(CH_2)_3\overset{\oplus}{N}Me_3 \quad X^{\ominus}$$

(**17**)

$$C_6H_{11}N{=}C{=}N{-}CH(CH_3){-}C_5H_4N$$

(**18**)

Typical of the generally reduced reactivity of aromatic carbodiimides towards acid-catalyzed reactions (*40*) di-*p*-tolylcarbodiimide is much inferior to DCC in oxidation reactions. Recently *N*-cyclohexyl-*N'*-*p*-toluenesulfonylcarbodiimide has been prepared and has been shown to promote the oxidation of benzhydrol in DMSO even without the addition of an acid (*45*). The general instability of this carbodiimide in the presence of DMSO, however, severely limits its usefulness. Also, several types of compounds that mechanistically resemble carbodiimides in their reactions have been tested and have been shown to lead to oxidation with less efficiency than DCC. Thus *N*-ethyl-5-phenylisoxazolium-3'-sulfonate (*46*), trichloroacetonitrile (*47*), and ethoxyacetylene (*48*) have been shown (*20*) to support the oxidation of testosterone to some degree. The reactions, however, were usually incomplete and frequently had to be heated. Oxidation of testosterone has also been achieved in low yield by Albright and Goldman (*30*) using ethoxyacetylene, and successful oxidation of the primary hydroxyl group of a nucleoside has been achieved by Harman et al. (*29*) using diphenylketen-*p*-tolimine and phosphoric acid in DMSO. Activation of DMSO by reaction with acid anhydrides will be considered in Section III.

In summary, the alcohol to be oxidized is dissolved in a 10–50% mixture of DMSO in benzene or ethyl acetate containing three molar equivalents of DCC. Following addition of 0.5 molar equivalents of H_3PO_4, dichloroacetic

acid, or pyridinium trifluoroacetate the stoppered reaction mixture is kept at room temperature, the length of reaction depending upon the choice of acid. The reaction mixture is diluted with ether or ethyl acetate and 3 molar equivalents of oxalic acid are added. After the gas evolution ceases, water is added, the dicyclohexylurea is removed by filtration, and the organic phase is extracted several times with water to remove the DMSO. The organic phase is then subjected to an appropriate work-up procedure. In the case of water-soluble compounds it is necessary to evaporate the aqueous extracts under high vacuum (or by lyophilization) to remove the DMSO.

B. Mechanistic Considerations

1. The Oxidation Reaction

In our original full paper on the DMSO–DCC oxidation reaction (*20*), we proposed a general mechanism that is summarized by Eqs. (2)–(4).

$$C_6H_{11}N{=}C{=}NC_6H_{11} + (CH_3)_2SO \xrightleftharpoons{H^{\oplus}} C_6H_{11}N{=}C(-NHC_6H_{11})-O-{}^{\oplus}S-(CH_3)_2 \quad \mathbf{(19)} \qquad (2)$$

$$\mathbf{(19)}\cdot H^{\oplus} + R^1R^2CHOH \xrightarrow{-H^+} R^1R^2CH-O-{}^{\oplus}S(CH_3)_2 \; \mathbf{(20)} + C_6H_{11}NHC(=O)NHC_6H_{11} \qquad (3)$$

$$R^1R^2CH-O-{}^{\oplus}S(CH_3)_2 \; \mathbf{(20)} \xrightarrow{-H^+} R^1R^2CH-O-{}^{\oplus}S(CH_3)-{}^{\ominus}CH_2 \; \mathbf{(21)} \longrightarrow R^1R^2C{=}O \; \mathbf{(22)} + CH_3SCH_3 \qquad (4)$$

In this mechanism the first step involves the acid-catalyzed addition of DMSO to DCC giving the sulfonium pseudourea (**19**). There is ample precedence for the addition of nucleophiles to protonated carbodiimides (*40,41*), and convincing evidence for the closely related addition of DMSO to ketenimines has been presented by Lillien (*49*). Also, nucleophilic displacements on oxysulfonium compounds [Eq. (3)] are well known and have been studied in detail by Johnson et al. (*50*). In the present case the separation of highly insoluble dicyclohexylurea renders this reaction virtually irreversible and provides a powerful driving force. The resulting formation of the alkoxysulfonium derivative (**20**) provides a mechanistic link between this reaction and other types of oxidations involving DMSO. These include the Kornblum oxidation of tosylates and alkyl halides (*6–10*) and the oxidative reactions of DMSO with chloroformates (*15*), epoxides (*13,14*), and diazonium salts (*16*). The final stage [Eq. (4)] in the sequence requires proton abstraction from the carbinol carbon of (**20**) and collapse of the intermediate carbanion into the carbonyl compound (**22**) and dimethyl sulfide. Since the reaction occurs under acidic conditions (e.g., phosphoric acid) or essentially neutral conditions (e.g., pyridinium trifluoroacetate usually in the presence of excess pyridine), direct bimolecular abstraction of a relatively unactivated proton seems unlikely and the results are best explained via a two-step mechanism. It is well known that protons on carbon directly attached to positively charged sulfur undergo ready exchange with D_2O (*51*) and that such an exchange is enhanced by several powers of ten in alkoxysulfonium salts such as (**20**) (*52*). Since such a proton abstraction leading to the formation of a highly sulfur d-orbital stabilized ylid (*53*) is known to proceed with bases as weak as acetate anion in, for example, the Pummerer reaction (*54,55*), a similar process promoted by phosphate anion or pyridine does not seem unlikely. The resulting ylid (**21**) is then ideally suited for an intramolecular proton abstraction via a five-membered cyclic transition state, followed by collapse to the observed carbonyl compound (**22**) and dimethyl sulfide. We have attempted to verify this mechanism by several experiments using isotopically labeled compounds (*56*). The initial formation of the DMSO–DCC adduct [Eq. (2)] seems on safe ground since oxidation of *p*-nitrobenzyl alcohol using ^{18}O-labeled DMSO together with DCC and anhydrous phosphoric acid leads to the formation of unlabeled *p*-nitrobenzaldehyde and dicyclohexylurea with essentially the same ^{18}O content as the original DMSO. This same conclusion has been reached indirectly by Albright and Goldman (*30*), who have oxidized ^{18}O-benzhydrol and have shown that the resulting benzophenone retained the label while the dicyclohexylurea was free of isotope. Both types of experiment effectively rule out a possible alternative mechanism for the oxidation reaction shown by Eqs. (5) and (6).

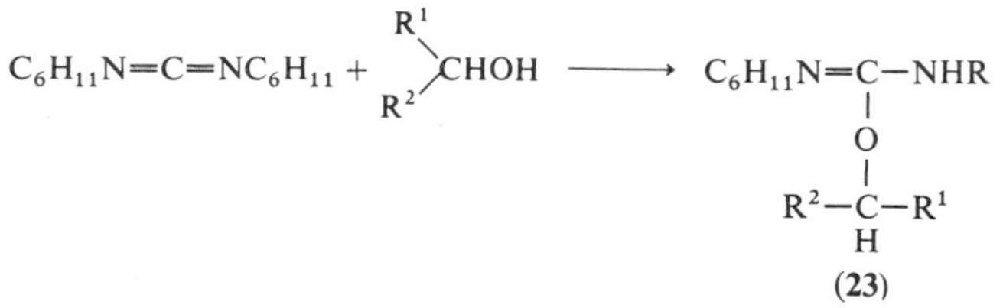

$$C_6H_{11}N{=}C{=}NC_6H_{11} + \begin{matrix} R^1 \\ R^2 \end{matrix}\!\!>\!CHOH \longrightarrow C_6H_{11}N{=}\underset{\substack{| \\ O \\ | \\ R^2-C-R^1 \\ H}}{C}-NHR \qquad (5)$$

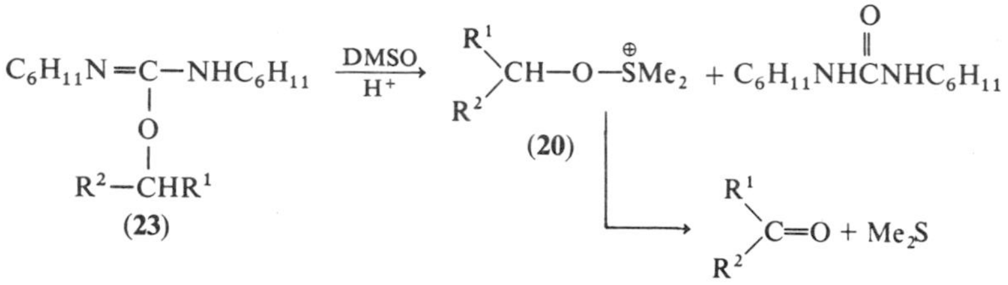

(23)

$$C_6H_{11}N{=}\underset{\substack{| \\ O \\ | \\ R^2-CHR^1}}{C}-NHC_6H_{11} \xrightarrow[H^+]{DMSO} \begin{matrix} R^1 \\ R^2 \end{matrix}\!\!>\!CH-O-\overset{\oplus}{S}Me_2 + C_6H_{11}NH\overset{O}{\overset{\|}{C}}NHC_6H_{11} \qquad (6)$$

(23) (20)

$$(20) \longrightarrow \begin{matrix} R^1 \\ R^2 \end{matrix}\!\!>\!C{=}O + Me_2S$$

In such a mechanism the final carbonyl oxygen should originate from DMSO and the oxygen of the dicyclohexylurea should originate from the starting alcohol. This pathway was also excluded by preparation of the proposed intermediate 1,3-dicyclohexyl-*O*-*p*-nitrobenzylpseudourea [(**23**), $R^1 = p\text{-}NO_2C_6H_4CH_2$—, $R^2 = H$], which was shown to give no *p*-nitrobenzaldehyde when reacted with DMSO, DCC, and anhydrous phosphoric acid (*20,30*).

Although the above isotope experiments demonstrate unequivocally that a DMSO–DCC adduct such as (**19**) is formed during the oxidation reaction, as yet we have been unable to demonstrate its accumulation. We have, for example, shown that the proton nmr spectrum of a mixture of DMSO and diisopropylcarbodiimide (1 : 2) in deuteriochloroform remains essentially unchanged upon addition of 0.5 molar equivalent of dichloroacetic acid (*57*). There is thus no observable accumulation of a sulfoxide–carbodiimide adduct since the dimethylsulfoxonium protons in such a compound would be expected to appear at about 3.3 ppm (*58*) rather than at the observed 2.58 ppm similar to DMSO itself. There was also no significant change in the resonances of the isopropyl protons of the carbodiimide other than the appearance of small amounts of the corresponding urea. Addition of an alcohol, however, led to immediate oxidation without detection of any intermediates. These results suggest that the acid-catalyzed addition of DMSO to a carbodiimide is reversible with the equilibrium lying far in favor of starting materials. Upon addition of an alcohol, however [Eq. (3)], the equilibrium is drawn to the right with simultaneous formation of the alkoxysulfonium derivative (**20**). Since accumulation of the latter should also be detected in the nmr spectrum, it must be assumed that ylid formation and collapse to products is rapid.

Other types of experiments also cast some light on the nature of these early steps. Thus it was shown (*59*) that the reaction of testosterone with DCC or diisopropylcarbodiimide, DMSO, and anhydrous *p*-toluenesulfonic acid in the presence of 2 equivalents of triethylamine relative to the acid leads to no observable oxidation. An identical reaction omitting only the triethylamine also gives no oxidation and is thus in agreement with our earlier observations on reactions catalyzed by strong acids. If, however, 2 equivalents of triethylamine are added to the free acid reaction after 15 min, there is immediate oxidation to an extent of 60–70% within 10 min. These experiments clearly indicate that the acid-catalyzed step precedes the base-catalyzed step, and suggest the accumulation of an oxysulfonium intermediate (**19** or **20**) which only goes on to products upon addition of a suitable base. Once again we have been unable to detect these proposed intermediates by nmr spectroscopy, and further work will be required to clarify the situation.

While it is difficult to design experiments to confirm Eq. (3) in the general mechanism, the general nature of the proton abstraction step [Eq. (4)] can be examined. Thus oxidation of 1,1-dideuteriobutanol (**24**) led to the formation of 1-deuteriobutyraldehyde (**26**) and monodeuteriodimethyl sulfide (**27**), both of which were isolated by vapor-phase chromatography and examined by mass spectrometry and nmr spectroscopy.

$$RCD_2{-}OH \xrightarrow[\text{DCC, } H^+]{\text{DMSO}} R{-}CD_2{-}O\overset{\oplus}{S}Me_2 \xrightarrow{-H^+} R{-}CD_2{-}O{-}\overset{\oplus}{S}(CH_3){-}\overset{\ominus}{C}H_2 \longrightarrow RC(=O)D + CH_3SCH_2D$$

(24) $R = C_3H_7$ **(25)** **(26)** **(27)**

The incorporation of one atom of deuterium into the dimethyl sulfide was unequivocally proved by the presence of a molecular ion (relative abundance 100%) at *m/e* 63. However, there was also a smaller peak (41% relative abundance) at *m/e* 62 which could be either the molecular ion from nondeuterated dimethyl sulfide or an M-1 fragment from (**27**). These results suggest that at least 70% of the dimethyl sulfide formed in the reaction contains deuterium and strongly supports an intramolecular proton abstraction mechanism via the sulfonium ylid (**25**). Alternatively, oxidation of unlabeled *n*-butanol using d_6-DMSO gave nondeuterated butyraldehyde and pentadeuteriodimethyl sulfide which was characterized by an intense molecular ion (relative abundance 100%) at *m/e* 67. Only a low intensity (relative abundance 11%) peak was present at *m/e* 68 corresponding to d_6-dimethyl sulfide. Similar conclusions have been reached by Sweat and Epstein (*60*), who have oxidized 3-^{3}H-cholestanol with DMSO, DCC, and pyridinium trifluoroacetate. These workers have shown that the resulting dimethyl sulfide, which was isolated as its mercuric chloride complex, contained tritium but that its specific activity was only 25–30% that of the starting alcohol. This discrepancy has been explained by the formation of dimethyl sulfide by pathways other than those of the oxidation reaction, and is qualitatively confirmed by the formation of small amounts of d_6-dimethyl sulfide from d_6-DMSO mentioned above. Sweat and Epstein (*60*) have obtained similar results during oxidation of 3-^{3}H-cholestanol with DMSO–acetic anhydride, although in this case the yield of ketone was only 21% and the formation of the acetate ester and the thiomethoxymethyl ether of the starting alcohol became the predominant reactions.

While the isotope experiments outlined above leave little doubt as to the gross mechanism of the oxidation reaction, the intermediacy of free alkoxysulfonium ions such as (**20**) has been questioned by Torssell (*58,61*). In his studies, Torssell prepared several alkoxysulfonium derivatives by independent routes and isolated them as crystalline tetraphenylborate salts. Subsequent reaction, of, for example, isobutoxydimethylsulfonium tetraphenylborate with DMSO, DCC, and pyridinium trifluoroacetate (100% excess pyridine) gave isobutanol and isobutyraldehyde in a ratio of 2 : 1, and oxidation of free isobutanol to isobutyraldehyde was complete under comparable conditions. The yield of aldehyde isolated as its dinitrophenylhydrazone derivative starting from the alkoxysulfonium tetraphenylborate was only about 10%. Attempts to isolate an oxysulfonium intermediate during the oxidation of several alcohols by the DMSO–DCC method were unsuccessful, and on the basis of these observations Torssell has rejected the intermediacy of a free alkoxysulfonium ion (**20**) during these reactions. The use in these studies of the tetraphenylborate salt of the alkoxysulfonium compound, however, raises some doubt as to the validity of these conclusions. Our earlier work has shown that strong acids themselves do not support oxidation, and that while the pyridine salts of these acids do function, the rates and extent of these reactions are markedly reduced. Since tetraphenylboric acid, while unknown as a free compound, is presumed to be a strong acid (*62*), there is serious reason to question whether the oxidation reaction would be expected to proceed in a satisfactory way under the conditions described by Torssell. In a similar way

Torssell has excluded the intermediacy of a free alkoxysulfonium intermediate in DMSO–acetic anhydride oxidations since treatment of isobutoxydimethylsulfonium tetraphenylborate with DMSO and acetic anhydride failed to yield isobutyraldehyde. This conclusion has been criticized by both Albright and Goldman (*30*) and by Johnson and Phillips (*55*), since under these conditions, acetate anion, which serves as the base for proton abstraction, is not present.

While questioning the free existence of an alkoxysulfonium intermediate, Torssell in general has supported most other details of our proposed mechanism. To avoid the intervention of a free alkoxysulfonium compound he has proposed an alternate three-body intermediate shown as (**28**).

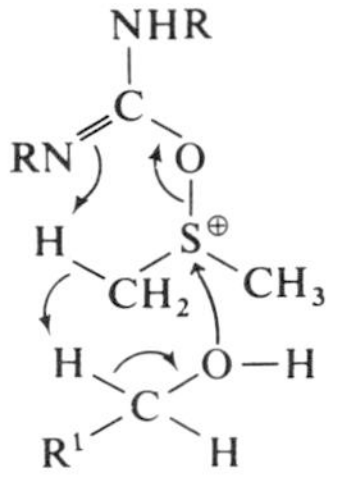

(28)

In this mechanism, attack by the alcohol upon the DMSO–DCC adduct initiates abstraction of an *S*-methyl proton by the nitrogen of the incipient dicyclohexylurea. As written, nucleophilic attack by the alcohol, subsequent proton abstraction, and collapse to products is considered to be a concerted process and has been criticized by Capon et al. (*63*) on electronic grounds.

The Torssell mechanism has, however, brought to light some observations that require us to modify our original mechanism. Oxidation of testosterone using d_6-DMSO and dichloroacetic acid was shown to be complete within 10 min. At this point the crystalline dicyclohexylurea was removed by filtration without addition of any protic solvents and was carefully washed with benzene. Mass spectrometry of this directly obtained material indicated the presence of 50% (±5%) of monodeuteriodicyclohexylurea and a small amount (less than 10%) of a dideuterio species. Since the amount of urea isolated was in excess of the theoretical 1 equivalent, presumably due to side reactions between the DCC and acid, the incorporation of deuterium probably approaches 1 mole per mole of alcohol oxidized. As a control the small amount of dicyclohexylurea that slowly crystallizes from a comparable reaction without addition of testosterone was also examined and was shown to contain less than 10% of the monodeuterio compound. This fact would appear to rule out any extensive deuterium exchange, and indicates a direct transfer of deuterium from DMSO to dicyclohexylurea during oxidation. Transfer of deuterium from d_6-DMSO into dicyclohexylurea has also been reported by Harmon and Zenarosa (*64a*) but has subsequently been refuted by the same authors (*64b*). This transfer appears to be best accomodated by a cyclic proton abstraction mechanism leading directly to the formation of the alkoxysulfonium ylid depicted in Eq. (7).

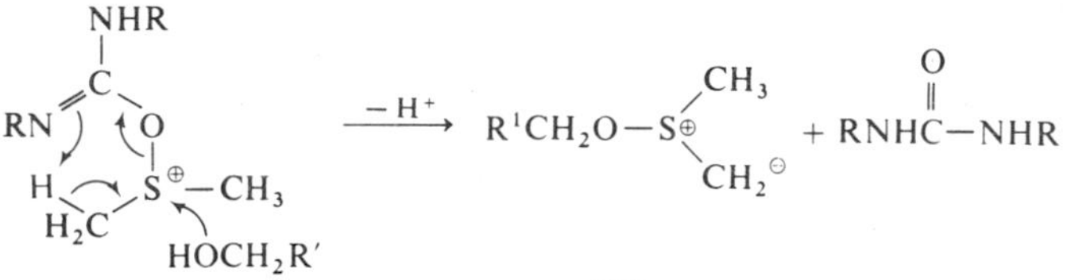

$$\text{cyclic adduct (RN=C(NHR)-O-}\overset{\oplus}{\text{S}}\text{(CH}_3\text{)CH}_2\text{-H} + \text{HOCH}_2\text{R}') \xrightarrow{-H^+} R^1CH_2O-\overset{\oplus}{S}(CH_3)\overset{\ominus}{C}H_2 + RNHC(=O)-NHR \quad (7)$$

(**21**)

The process could either be concerted, as shown in Eq. (7), or could proceed by initial formation of a tetracovalent sulfur intermediate (**29**) which subsequently undergoes collapse to the ylid via a cyclic process, as shown in Eq. (8).

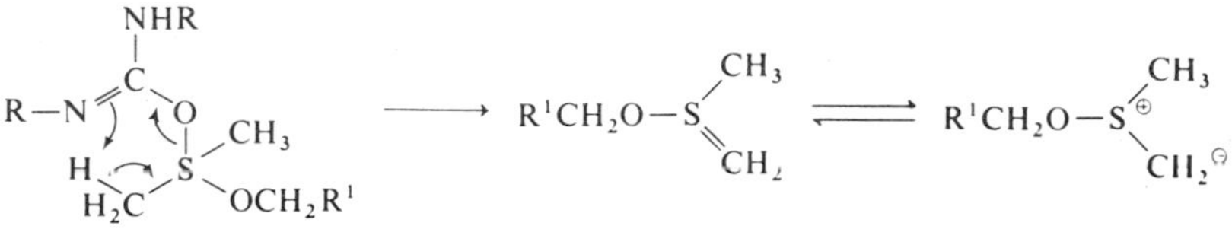

$$\text{(29)} \longrightarrow R^1CH_2O-S(CH_3)=CH_2 \;\text{(21a)} \rightleftharpoons R^1CH_2O-\overset{\oplus}{S}(CH_3)\overset{\ominus}{C}H_2 \;\text{(21b)} \quad (8)$$

(**29**) (**21a**) (**21b**)

The latter possibility is attractive since as long as a positive charge remains on sulfur in the DMSO–DCC adduct (**19**), its inductive effect could favor dissociation into DMSO and protonated DCC as suggested by our nmr studies in the absence of an alcohol. Once the charge is removed, however, by addition of the alkoxy group [as in (**29**)], the electron flow could be reversed and would then lead rapidly to the alkoxysulfonium ylid (**21**) and thence to the carbonyl compound (**29** → **22**).

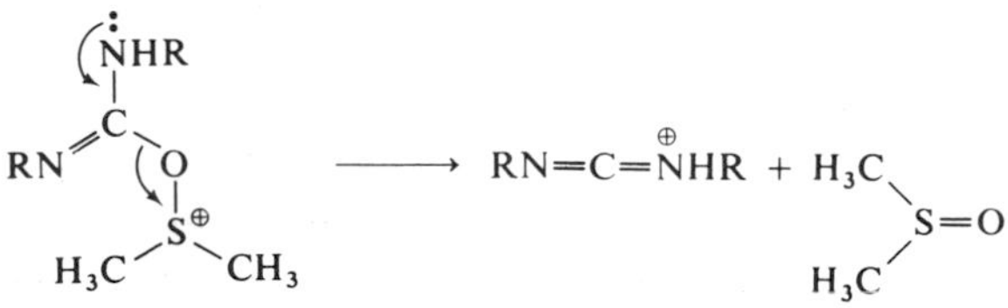

$$\text{(19)} \longrightarrow RN=C=\overset{\oplus}{N}HR + (H_3C)_2S=O$$

(**19**)

Since in the above mechanisms the proton abstraction steps leading to both the formation of ylid (**21**) and its collapse to products (**21** → **22**) are intramolecular, the formal requirement for an external base is lost. The failure of strong acids to support oxidation then can be explained by decreased nucleophilicity of the alcohol or by protonation of either the adduct (**29**) or the ylid (**21**). The modifications to our original mechanism that are suggested above appear to explain our experimental observations quite well. There is still a need, however, for further work to completely understand this interesting chain of events.

2. Thiomethoxymethyl Ether Formation

As previously mentioned, the oxidation of testosterone (**10**) to androst-4-ene-3,17-dione (**11**) was used in our original studies as a convenient model to determine optimal reaction conditions. With pyridinium trifluoroacetate as the proton source this reaction leads to quantitative conversion to the dione (**11**) as judged by thin-layer chromatography. The isolation of pure (**11**) could be achieved in 92% yield by direct crystallization. When anhydrous orthophosphoric acid was used as the proton source, (**11**) was still obtained in 87.5% yield, but, in addition, traces of two by-products were formed, the major one being the thiomethoxymethyl ether (**12**). This type of compound is a frequently encountered minor by-product arising during oxidation of relatively nonhindered alcohols. The formation of thiomethoxymethyl ethers always seems to be highest when phosphoric acid is used and can become a serious side reaction during oxidations using DMSO and acetic anhydride (Section III). It was originally suggested (*65*) that their formation could arise via either an intramolecular rearrangement of the oxysulfonium ylid intermediate (**30** → **31**) or by alkylation of the alcohol by the methyl methylenesulfonium ion (**32**).

$$\mathrm{RO}\!-\!\overset{\oplus}{\mathrm{S}}(\mathrm{CH_3})\!-\!\overset{\ominus}{\mathrm{C}}\mathrm{H_2} \longrightarrow \mathrm{ROCH_2SCH_3} \longleftarrow \mathrm{ROH} + \mathrm{CH_2}\!=\!\overset{\oplus}{\mathrm{S}}\mathrm{CH_3}$$

(30) **(31)** **(32)**

The latter ion could originate via dissociation of the oxysulfonium ylid (**30**) or by collapse of the DMSO–DCC adduct (**33**) or the related ylid (**34**).

$$\text{(33)} \longrightarrow CH_2{=}\overset{\oplus}{S}{-}CH_3 + C_6H_{11}NH{-}\overset{O}{\overset{\|}{C}}{-}NHC_6H_{11} \longleftarrow \text{(34)}$$

(33) (34)

Since the alkoxysulfonium ylid (**30**) is considered to be a common intermediate in the mechanisms of oxidations using both DMSO–DCC and DMSO–acetic anhydride (*30*), both methods would be expected to lead to the same relative proportions of carbonyl compound and thioether when applied to the same alcohol. Since, in fact, the formation of thiomethoxymethyl ethers is vastly more prevalent during DMSO–acetic anhydride reactions, the intermediacy of the sulfonium ion (**32**) seems to be strongly favored. This same ion also appears to be involved in a variety of reactions of DMSO and DCC with functional groups other than alcohols leading to the introduction of thiomethoxymethyl groups (*33*). The very minor extent of this side reaction using DMSO–DCC indicates that the formation of (**32**) via collapse of (**33**) or (**34**) is not a favored process, and this concept is supported by our other studies on reactions of DMSO and DCC with phenols (*34,35*). Thus we and others (*66*) have shown that the ion (**32**) formed by other routes does efficiently alkylate compounds such as anisole. Thiomethoxymethyl anisoles are not, however, formed from DMSO, DCC, and anisole in the presence of anhydrous phosphoric acid. The reason may well be that, as discussed earlier, there is no significant accumulation of (**33**) prior to reaction with a nucleophile. However, the formation of (**32**) by collapse of the acetoxydimethylsulfonium ion, which is undoubtedly the first intermediate in reactions of DMSO and acetic anhydride, is facile and constitutes a key step in the Pummerer rearrangement (*54,55*).

C. Scope of the Oxidation Reaction

In recent years the DMSO–DCC method has been used by many workers for the oxidation of a wide variety of alcohols. The DMSO–DCC reagent provides a powerful oxidizing medium capable of oxidizing rather inert alcohols while operating under very mild conditions suitable for use with sensitive molecules.

1. Applications to Steroids

Oxidations of hydroxyl groups at various positions in steroid molecules have been examined and, in general, proceed well. As previously discussed, the oxidation of testosterone (**10**) to (**11**) has been extensively studied as a model reaction and can be conducted with an isolated yield of 92% (*65*). Other simple 17-alcohols such as 3-methoxyestra-1,3,5,8-tetraen-17-ol and its 8(14) seco derivative have also been oxidized in very high yields (*67*) as have isolated hydroxyl groups in other positions, e.g., cholestanol (*60,65*).

To ascertain any steric requirements, several pairs of epimeric steroidal alcohols have been oxidized and the relative rates have been followed by quantitative thin-layer chromatography (*65*). In the case of largely unhindered alcohols, only minor differences in rate were observed. Thus, the rates with testosterone (**10**) and its 17α-hydroxy epimer (epitestosterone) were very similar, especially using anhydrous phosphoric acid. With pyridinium phosphate, the overall rate was considerably slower and there appeared to be a reduced rate with the 17α-isomer. In all cases, the same final product (**11**) was formed in greater than 90% yield. Similar minor differences in rate were found between the equatorial hydroxyl group of 3β-hydroxy-5-β-pregn-16-en-20-one (**35a**) and its axial 3α-hydroxy epimer (**35b**).

More dramatic steric differences were found in the cases of the much more hindered 11-hydroxy steroids. Thus, the equatorial hydroxyl group of 11α-hydroxyprogesterone (**36a**) was smoothly and quantitatively oxidized to the 11-ketone (**36c**) by using either anhydrous phosphoric acid or pyridinium trifluoroacetate. On the other hand, the epimeric axial 11β-hydroxyprogesterone (**36b**) was completely inert under comparable conditions using pyridinium trifluoroacetate, and use of phosphoric acid gave only 11% of the 11-keto compound (**36c**), and 20% of the dehydration product pregna-4,9(11)-diene-3,20-dione (**37a**). In a very similar way the axial 11β-hydroxyl group of corticosterone-21-acetate (**36d**) was inert to oxidation using pyridinium trifluoroacetate and gave only 23% of the dehydration product 9(11)-dehydrocorticosterone-21-acetate using phosphoric acid. Simultaneous oxidation of both the 11α- and 17β-hydroxyls of (**38**) to the corresponding dione in 80% yield has been reported by Turnbull et al. (*37*). However, 11β,17α,20α,21-tetrahydroxypregn-4-en-3-one-21-acetate was selectively oxidized to the 20-ketone (*32*).

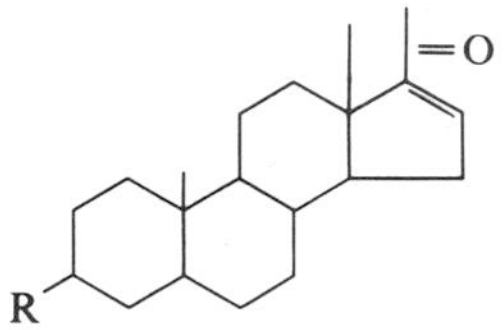

(35)(a) R = β-OH
(b) R = α-OH

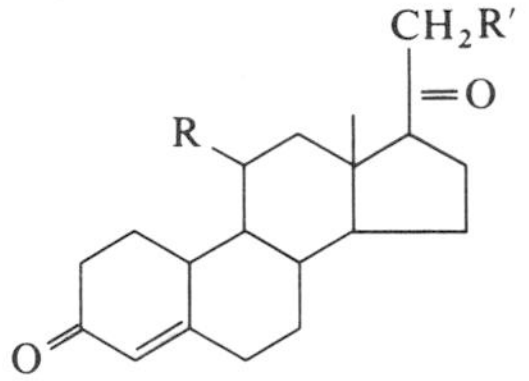

(36)(a) R = α-OH, R′ = H
(b) R = β-OH, R′ = H
(c) R = O, R′ = H
(d) R = β-OH, R′=OAc
(e) R = α-OCH_2SCH_3, R′=H

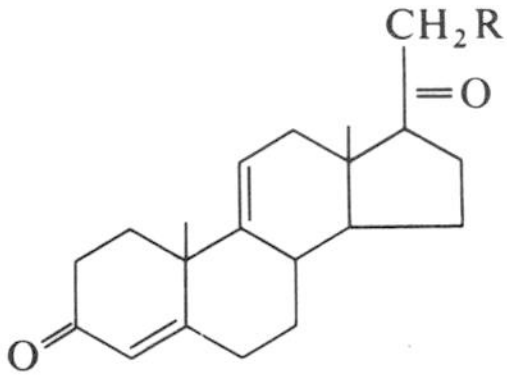

(37)(a) R = H
(b) R = OAc

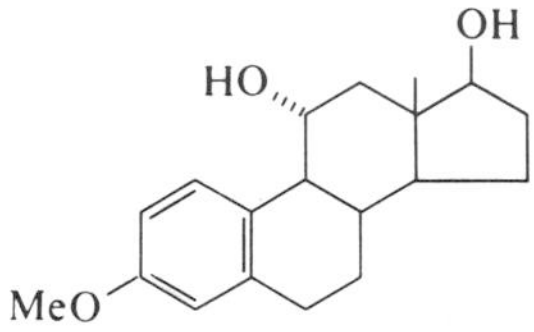

(38)

The observed greater ease of oxidation of the equatorial 11α-hydroxysteroids relative to their axial hydroxy counterparts is in marked contrast to what is known to occur during oxidations using chromic acid (*68*). There the axial 11β-hydroxyl groups are much more readily oxidized and this has been explained (*69*) by the greater steric accessibility of the 11α-hydrogen which must be abstracted from the intermediate chromate ester during the rate-limiting step. The effect is perhaps more correctly explained by relief of the rather high ground state energy in going from the hindered alcohol to the transition state and product (*68*). Since in the DMSO–DCC oxidation reaction the most sterically demanding, and probably rate-limiting, step is attack of the alcohol upon the DMSO–DCC adduct, the equatorial alcohol would be expected to be oxidized most rapidly. It is interesting to note that, in contrast to the above, oxidation of the axial 11β-hydroxyl group of hydrocortisone-21-acetate proceeds smoothly, although slowly, using DMSO–acetic anhydride. This is presumably due to the much smaller bulk of the intermediate acetoxyidmethylsulfonium ion relative to the DMSO–DCC adduct (**19**), thus permitting more ready attack by the rather hindered axial alcohol. On the other hand, oxidation of equatorial 11α-hydroxysteroids using DMSO–acetic anhydride gives predominantly the corresponding thiomethoxymethyl ethers (*30*).

Because of the mildness of the oxidation reaction it is possible to effect some reactions that are otherwise troublesome. For example, oxidation of homoallylic alcohols such as Δ^5-3-hydroxy-steroids to the corresponding β-γ-unsaturated ketones has been a source of some difficulty owing to the ready isomerization of the final products to the conjugated 4-en-3-ones under either acidic or basic conditions. In the past, this type of oxidation has been achieved by a three step process involving bromination and debromination of the olefin (*70*) or by precisely controlled treatment with chromic acid in acetone (*71*). While minor isomerization of the double bond into conjugation did occur during oxidation of androst-5-en-3β-ol-17-one (**39**) using DMSO–DCC and anhydrous phosphoric acid, comparable reactions using pyridinium trifluoroacetate contained almost none of the 4-en-3-one (**11**), and the desired β-γ-unsaturated ketone (**40**) was obtained in pure form and in 70% yield by direct crystallization from the extracted reaction mixture. This method has been used with comparable success for the oxidation of a variety of related 3β-hydroxy-androst-5-en derivatives (*38,65,72*) and of cholesterol (*38,65*).

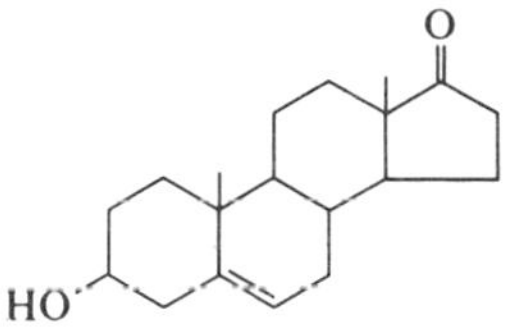

(**39**) 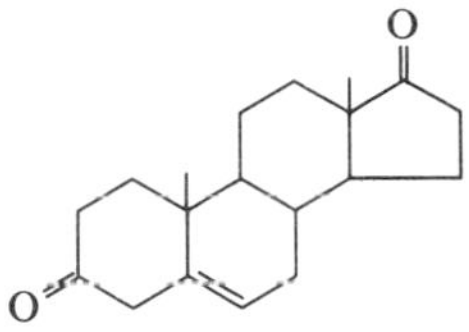 (**40**)

While the major uses of the DMSO–DCC reaction to date have been in the oxidation of secondary alcohols to ketones, perhaps the most potentially valuable applications lie in the preparation of aldehydes from primary alcohols. Although many methods are known for the synthesis of aldehydes (*73*), the direct oxidation of primary alcohols is a frequently troublesome laboratory procedure due mainly to a considerable tendency toward overoxidation to carboxylic acids. Since the DMSO–DCC method requires nucleophilic attack by a free hydroxyl group as a prerequisite to oxidation, an aldehyde, once formed, is mechanistically incapable of further reaction. In support of this idea we have carefully examined the reaction mixtures resulting from oxidation of many primary alcohols, and in no case have we been able to detect the formation of any carboxylic acids. Relatively few steroidal aldehydes have been prepared as yet by this method, but oxidations of, for example, cholane-24-ol (**41a**) and androst-4-en-19-ol-3,17-dione (**42a**) have given the corresponding crystalline aldehydes

(**41b** and **42b**) in yields of 85 and 82% (*65*). Similarly, 3β-acetoxyandrost-5-en-19-ol-17-one gave the 19-aldehyde in 53% yield (*65*), and a tritium-labeled derivative of 4,4,14-trimethylchol-8-en-24-ol was prepared for studies on the biosynthesis of lanosterol (*74*).

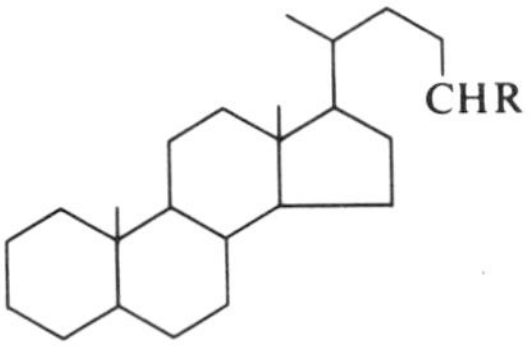

CHR

RCH

O

O

(**41**)(**a**) R = H, OH
(**b**) R = O

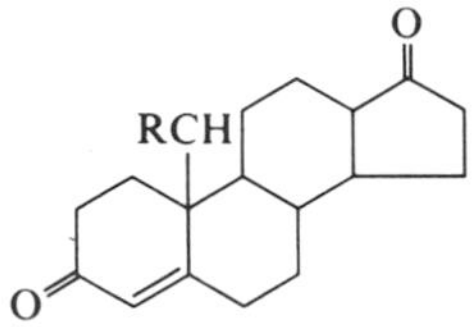

(**42**)(**a**) R = H, OH
(**b**) R = O

2. Applications to Carbohydrates

Few classes of organic compounds offer a range of hydroxyl functions comparable to the carbohydrates. In spite of its obvious synthetic importance, however, the oxidation of isolated hydroxyl groups in protected sugars frequently has proved to be a source of difficulty (*75*). The advent, during the past few years, of mild methods of oxidation using DMSO in the presence of DCC, acetic anhydride (*30*), or phosphorus pentoxide (*31*) and the use of ruthenium tetroxide (*76*) has led to a burst of research concerning the chemistry of sugar derivatives containing isolated carbonyl functions.

Numerous examples of oxidations have been described in carbohydrate chemistry using the DMSO–DCC method. In these reactions carbonyl functions have been introduced at all possible positions in simple pentose and hexose molecules, and the isolated yields are usually quite good. A majority of the examples have dealt with the synthesis of ketones but, as mentioned earlier, the specific oxidation of primary hydroxyl groups to the aldehyde level promises to be of particular value.

Our own particular interest in the preparation of sugar aldehydes originated with a desire to develop a practical synthetic route to nucleoside 5′-aldehydes (**44**, **45**). Such compounds constitute versatile intermediates in the synthesis of a variety of nucleoside analogs containing modified sugars, and one member of this series, adenosine 5′-aldehyde [(**45**); B = adenine], has been identified as a product resulting from the photolysis of vitamin B_{12} coenzyme (*77*). Oxidation of a large number of 2′,3′-*O*-substituted ribonucleosides and 3′-*O*-substituted-2′-deoxynucleosides has been investigated in this laboratory (*20,22,78*). Generally the conversion of the 5′-hydroxyl group of the starting material (**43**) to the corresponding aldehyde (**44**) has been shown by thin-layer chromatography to be very high with only trace amounts of unreacted starting materials and the corresponding thiomethoxymethyl ethers as contaminants. The crude products have been directly used in various condensation reactions, for example, with the phosphorane (**46**) leading to the vinyl phosphonates (**47**) which can be efficiently converted into 6′-deoxyhomonucleoside phosphonic acids (**48**) (*79*).

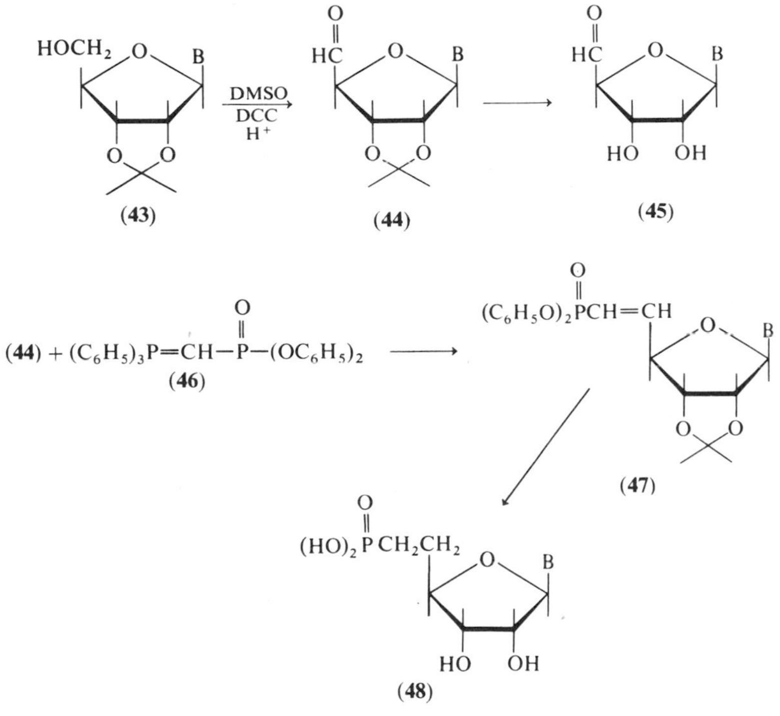

HOCH$_2$ O B; DMSO, DCC, H$^+$; HC(=O) O B; HO OH

(**43**) (**44**) (**45**)

$$(\mathbf{44}) + (C_6H_5)_3P{=}CH{-}\overset{O}{\overset{\|}{P}}{-}(OC_6H_5)_2 \longrightarrow (C_6H_5O)_2\overset{O}{\overset{\|}{P}}CH{=}CH\text{-}$$

(**46**) (**47**)

$(HO)_2\overset{O}{\overset{\|}{P}}CH_2CH_2$ O B; HO OH

(**48**)

Crystalline derivatives of (**44**) and (**45**) have been obtained, but isolation of the free aldehydes in a completely homogeneous form has proved to be difficult. The major problem has been a marked instability of the protected aldehydes [e.g., (**44**) or the related derivatives] towards any form of adsorption chromatography. Thus, attempted chromatography of, for example, 2′,3′-*O*-benzylideneuridine 5′-aldehyde (**49**) on silicic acid, alumina, florisil, etc., leads to rapid elimination of the acetal grouping giving the 3′,4′-unsaturated aldehyde (**50**) (*80*).

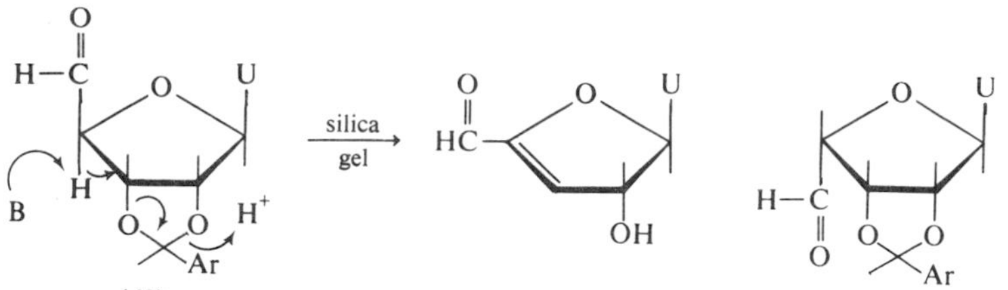

(49) (50) (51)

The ease of such eliminations is strongly affected by the nature of the acetal function, and similar reactions can be brought about by a variety of bases. Some epimerization to the corresponding α-L-lyxosyl nucleoside 5′-aldehyde (**51**) also accompanies attempted chromatography of the protected aldehydes. If chromatographic separation is avoided, reduction of the crude protected 5′-aldehydes (**44**) with sodium borotritiide provides a convenient route to specifically 5′-tritiated nucleosides (*81,82*). Very recently, however, pure nucleoside 5′-aldehydes have become available via isolation of crystalline *N,N*′-diphenylimidazolidine derivatives from which the carbonyl group can be regenerated under mild conditions (*78b*).

When chromatography can be avoided elimination reactions frequently cease to be a problem. Thus oxidation of methyl 2,3-*O*-isopropylidene-β-D-ribofuranoside (**52a**) using DMSO–DCC and phosphoric acid proceeds in high yield, and the resulting aldehyde (**52b**) can be isolated in 85% yield by sublimation (*78*). Attempted chromatographic isolation, however, once again leads to some elimination and epimerization. Since *N*-trifluoroacetyl- and *N*-dichloroacetyl-*N,N*′-dicyclohexylureas are volatile, we have found the use of anhydrous phosphoric acid to be preferable to the haloacetic acids in those cases where the product is to be purified by distillation. In a similar way, oxidation of 1,2;3,4-di-*O*-isopropylidene-β-D-galactopyranose (**53a**) has been successfully oxidized to the 6-aldehydo compound (**53b**), which was isolated by distillation in up to 83% yield (*78b,83–85*). In other cases, sugar aldehydes have been obtained in high yield when groups incapable of elimination are present (e.g., benzyl ethers as in (**54**)) (*86*), and when isolation is facilitated by formation of a bisulfite complex as with (**55**) (*87*).

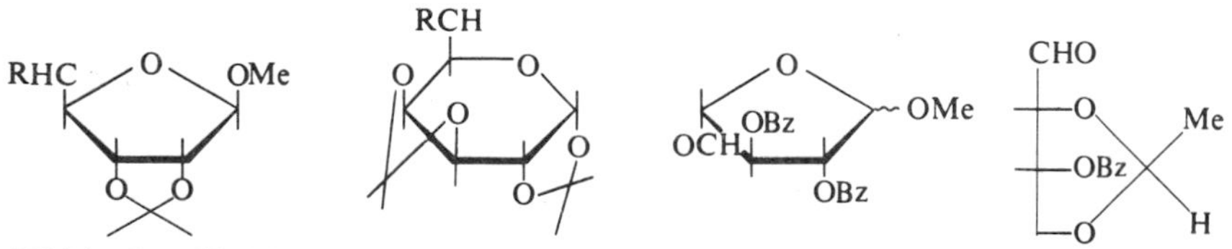

(**52**)(**a**) R = H, OH
(**b**) R = O

(**53**)(**a**) R = H, OH
(**b**) R = O

(**54**)(**a**) α-anomer
(**b**) β-anomer

(**55**)

In certain cases, however, elimination of acyl groups has been observed.

Thus, oxidation of both the α and β forms of methyl 2,3,4-tri-*O*-acetyl-D-glucopyranoside (**56a**) gives the simple 6-aldehydes (**56b**) and the unsaturated aldehydes (**57**) arising by elimination of acetate (*88*). The relative proportions of both types of product vary considerably, with the β-anomer being considerably more susceptible to elimination. Oxidation of the α-anomer has indeed been reported in quantitative yield without elimination (*89*). At least in part the loss of acetate occurs during chromatography, but (**57**) also appears to be present in the crude reaction mixture. Oxidation of 1,2,3,4-tetra-*O*-acetyl-β-D-glucopyranose also gives the unsaturated aldehyde as the major isolated product (*88*), although preparation of the simple aldehyde has been claimed without experimental detail (*90*). Interestingly enough, oxidation of several 4-*O*-sulfonyl glucose derivatives [(**58**) (*88*) and (**59**) (*91*)] has given the 4-sulfonyl-6-aldehydes apparently without extensive elimination. Further work is necessary to clarify the synthetic utility of these reactions. When oxidation of similar sugars is done using the more strongly basic DMSO–pyridine–sulfur trioxide reagent in the presence of triethylamine, the formation of unsaturated aldehydes becomes an efficient process (*92*).

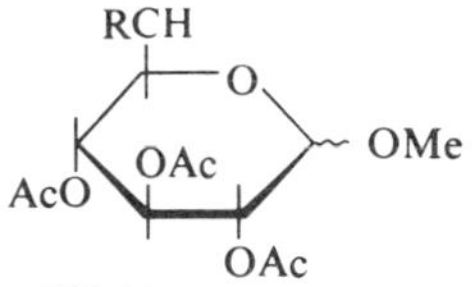

(**56**)(**a**) R = H, OH
(**b**) R = O

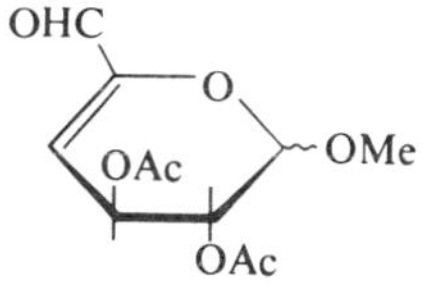

(**57**)

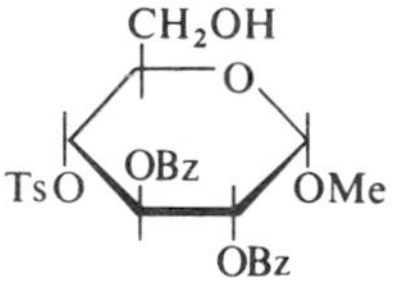

(**58**)

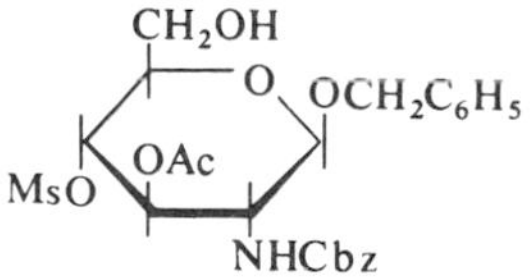

(**59**)

Applications to the oxidation of secondary hydroxyl groups in carbohydrates are more common, and side reactions are less frequently encountered. The oxidation of hydroxyl groups in a wide variety of environments has been successfully achieved, and in general the yields are good. Acyclic hydroxyl groups such as that in methyl 6-deoxy-2,3-*O*-isopropylidene-β-D-allofuranoside (**60**) (*93*) are readily oxidized to the corresponding ketones, and this type of reaction has been applied to the conversion of nonreducing sugar alcohols such as 2,3,5-tri-*O*-benzoyl-1-*O*-trityl-D-arabinitol (**61**) (*94*) and 1,2;3,4-di-*O*-isopropylidene-L-rhamnitol (**62**) (*93*) into the corresponding open-chain ketoses.

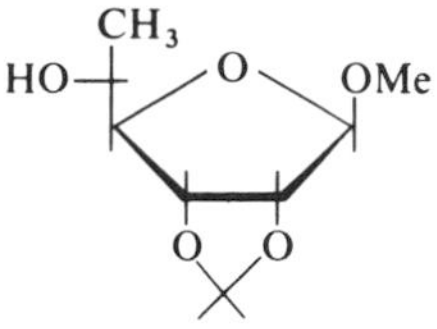

(60) 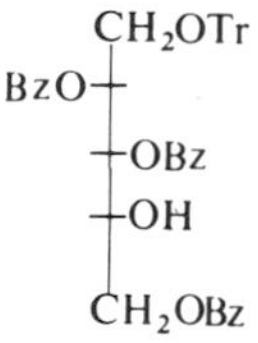 (61) 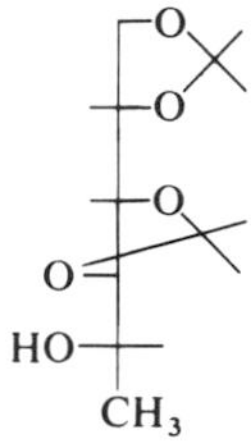 (62)

A much wider variety of secondary hydroxyl groups situated on five- or six-membered rings has been examined, and several interesting observations have been noted regarding the influence of adjacent functional groups. Adjacent acetal functions certainly appear to have little deleterious effect upon the reaction. Oxidation of, for example, 1,2;3,4-di-*O*-isopropylidene-6-*O*-methyl-*epi*-inositol (**63**) (*95*) and 6-deoxy-1,2;3,4-di-*O*-isopropylidene-*cis*-inositol (**64**) (*96*) gave the corresponding ketones in yields of 85 and 82%, respectively, and oxidation of 5 deoxy-1,2-*O*-isopropylidene-*β*-L-arabinofuranose (**65**) was a key step in the synthesis of streptose by Dyer et al. (*97*).

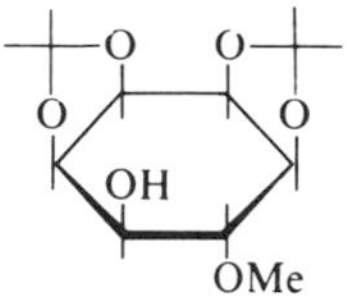

(63) 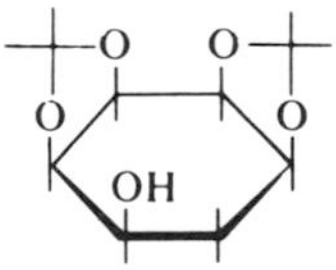 (64) 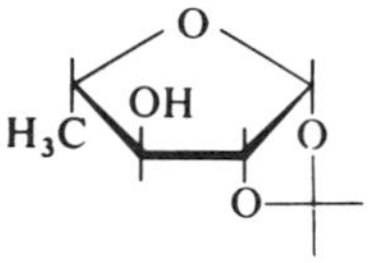 (65)

Baker and Buss (*98*) have, however, reported that certain hydroxyl groups flanked by acetal and ether functions do not undergo oxidation. This report was based upon their failure to observe a carbonyl band in the infrared spectra of the crude products after oxidation of 1,2;5,6-di-*O*-isopropylidene-*α*-D-glucofuranose (**66**) and methyl 3,4-*O*-isopropylidene-*β*-L-arabinopyranoside (**67**). We have, however, found that oxidation of (**66**) using the DMSO–DCC method proceeds almost quantitatively as judged by vapour-phase chromatography and gives the desired 3-keto sugar (**68**) as its crystalline hydrate in yields of up to 92% (*88*). Also, oxidation of the optical enantiomer of (**67**) has been achieved by Novak and Šorm (*99*), although the isolated yield was apparently not as high as that obtained with chromic oxide in pyridine. Apparently the failure to detect oxidation of (**66**) and (**67**) was a consequence of the great tendency of keto sugars such as (**68**) to form stable hydrates which show no carbonyl band in their infrared spectra. It is also difficult to detect oxidation of

(**66**) by thin-layer chromatography since the hydrated product and starting material have very similar mobilities. However, the oxidation can be followed readily by vapor-phase chromatography.

The oxidation of 1,2;4,5-di-*O*-isopropylidene-β-D-fructopyranose (**69**) to the related crystalline 3-keto compound also has been achieved in 69% yield using DMSO–DCC–phosphoric acid (*100*). This same oxidation also has been accomplished with comparable yields using DMSO–acetic anhydride (*101–103*), and borohydride reduction of the product provides a facile synthesis of psicose derivatives. In our hands, the DMSO–DCC method provides the most ready route to the pure product.

The very successful oxidations of (**66**) in yields of 60–92%, by using DMSO in conjunction with DCC (*88*), acetic anhydride (*104*), or phosphorus pentoxide (*31*), point out the powerful nature of these reagents. Oxidations of (**66**) with a great variety of reagents have been attempted over the years, and until the use of ruthenium tetroxide (*76*) and the DMSO-based oxidants became known, the maximum yield of (**68**) that had been achieved was 6% using chromic oxide in pyridine–acetic acid (*105*).

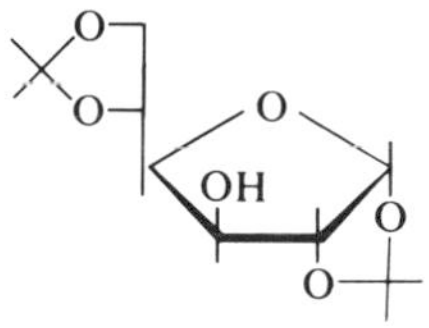

OH O O O O

(**66**)

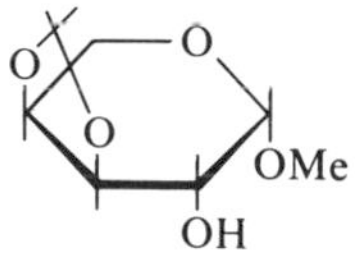

O O O OMe OH

(**67**)

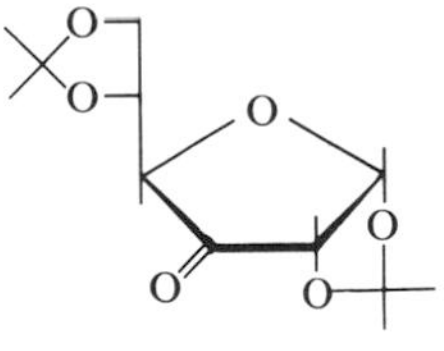

O O O O O O

(**68**)

O O HO O CH_2O O

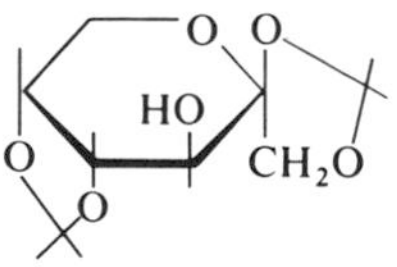

(**69**)

The presence of adjacent sulfonate esters in an open-chain sugar, or of an adjacent equatorial sulfonate ester on a pyranose ring, appears to present no problems. Thus, Baker and Buss (*98*) have shown that oxidations of 1,2;5,6-di-*O*-isopropylidene-3-*O*-mesyl-D-mannitol (**70**) and of methyl 4,6-*O*-benzylidene-2-*O*-tosyl-α-D-glucopyranoside (**71**) give the corresponding keto sugars in yields of 74 and 80%, respectively. As yet, no examples of the oxidation of hydroxyl groups with axial sulfonate neighbors have appeared.

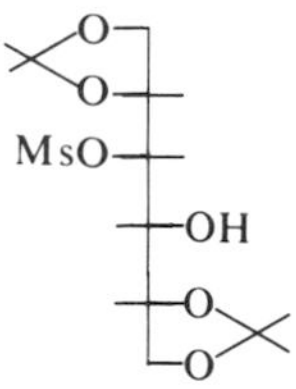

(70) 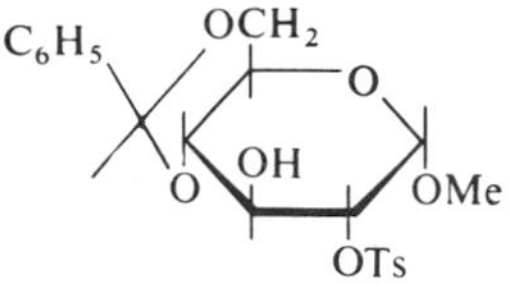 (71)

The presence of adjacent benzamido functions however, can lead to epimerization (*106*). Thus, the equatorial benzamido compounds (**72**) and (**75**) are oxidized to the corresponding ketones (**73**) and (**76**) in yields of 96 and 95%, respectively. The identical ketones, however, are also obtained in high yield starting with the axial benzamido compounds (**74**) and (77). As expected, (73) is also obtained from the *manno* epimer of (**72**).

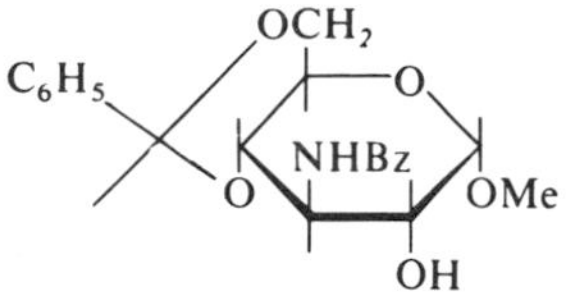

(72) 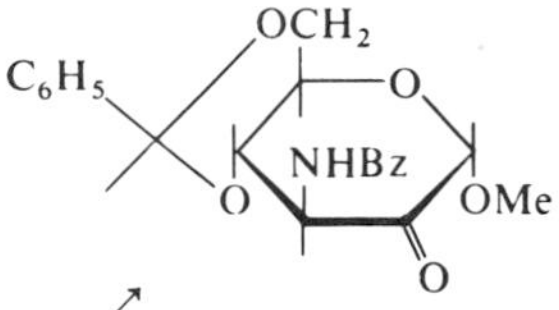 (73)

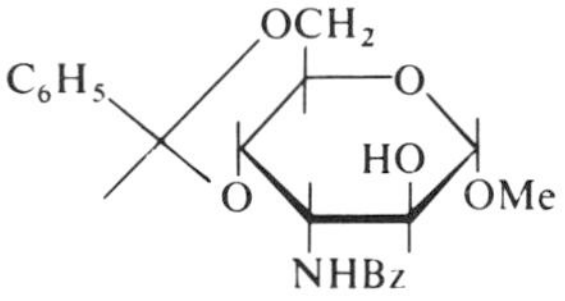

(74)

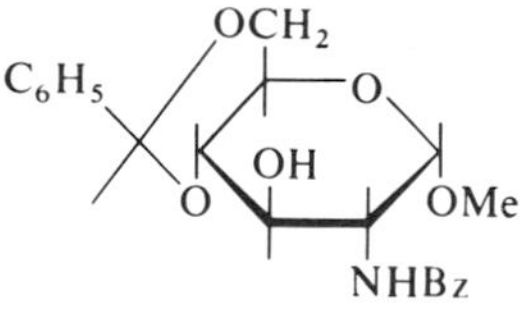

(75) 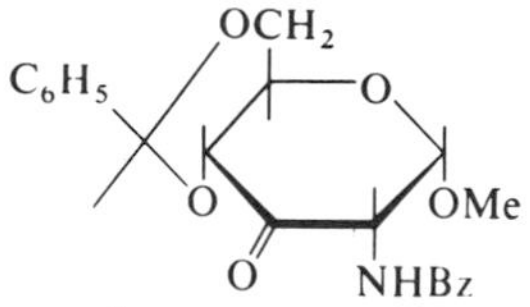 (76)

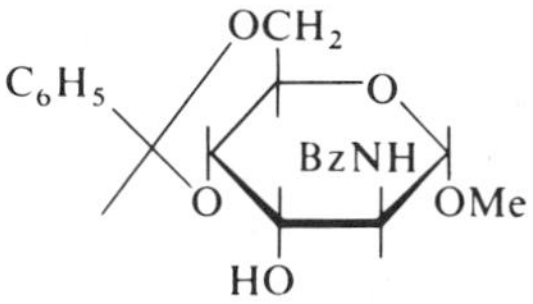

(77)

It remains to be seen whether epimerization can be avoided by using less acidic catalysts such as pyridinium trifluoroacetate or pyridinium phosphate. The related epimerization of an axial azido function during oxidation of (**78**) with DMSO–acetic anhydride has also been reported (*107*) and has been interpreted as an acid-catalyzed reaction subsequent to oxidation. Partial epimerization of the 2-benzyloxy group during oxidation of benzyl 2,3,5,6-tetra-*O*-benzyl-L-gulonate (**79**) using DMSO–DCC pyridinium trifluoroacetate has been noted (*108*) and is not surprising in view of the presumed acidity of the resulting *β*-keto ester. The only recorded case of elimination accompanying oxidation of a secondary alcohol involves the formation, in 50% yield, of the unsaturated keto sugar (**81**) during oxidation of methyl 3-acetamido-4,6-di-*O*-acetyl-3-deoxy-α-D-mannopyranoside (**80**) (*109*). As in the cases of other eliminations observed during the oxidation of primary alcohols with DMSO–DCC (*88*) and of primary and secondary alcohols with DMSO–SO_3-pyridine (*92*), the observed reaction appears to be a *cis* elimination, and there is little reason to believe that prior epimerization of the 3-acetamido function of (**80**) precedes loss of acetate. Oxidation of the closely related methyl 3-benzamido-4,6-*O*-benzylidene-3-deoxy-α-D-mannopyranoside gives a 70% yield of the saturated 2-ketone (*109*).

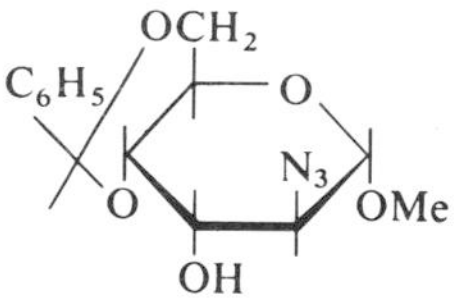

(**78**) 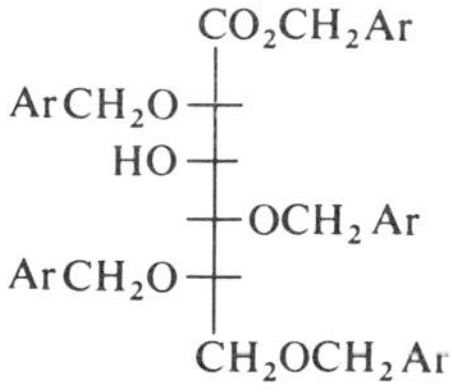 (**79**)

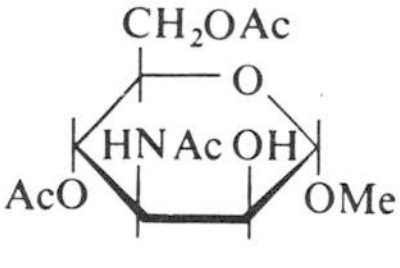

(**80**) 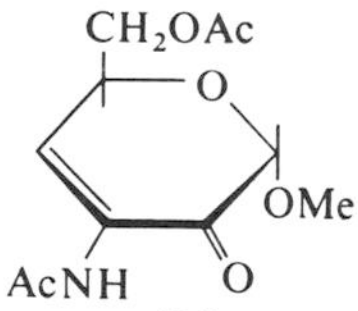 (**81**)

Only one example of the oxidation of the hemiacetal hydroxyl in a reducing sugar has been described using DMSO–DCC although several other noncarbohydrate compounds have been studied and will be considered later. Horton and Jewell (*110*) have shown that treatment of 2,3,5,6-di-*O*-isopropylidene-α-D-mannofuranose (**82**) gave the corresponding mannonolactone (**83**) in 35% yield.

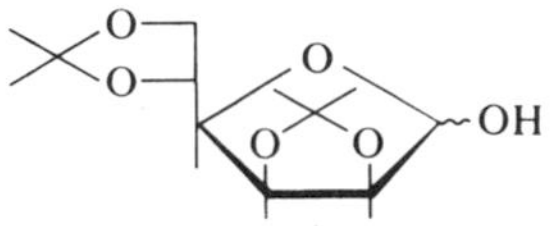

(82) 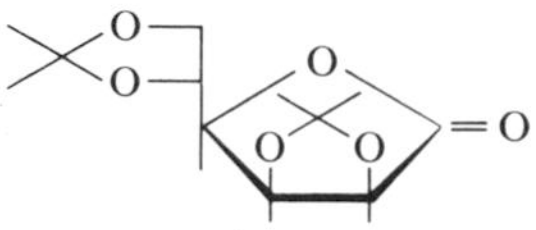 (83)

A number of other suitably substituted sugars such as 1,6-anhydro-2,3-*O*-isopropylidene-β-D-mannopyranose (*110*) and 3,6-dideoxy-1,2-*O*-isopropylidene-α-D-galactofuranose (*111*) also have been oxidized successfully as have several acyclic sugar derivatives (*112,113*) and anhydrohexitols (*114*).

The oxidation of secondary hydroxyl functions in nucleosides is complicated by the extreme alkaline lability of the ketonucleosides. As mentioned earlier, all attempts to oxidize the 3′-hydroxyl group of 5′-protected thymidine derivatives have led to spontaneous elimination of thymine with no observable accumulation of keto intermediates. However, the oxidation of 2′,5′-di-*O*-trityluridine (**84a**) and of 3′,5′-di-*O*-trityluridine (**86a**) with DMSO–DCC in the presence of pyridinium trifluoroacetate leads to the isolation of 2′,5′-di-*O*-trityl-3′-ketouridine (**85a**) and 3′,5′-di-*O*-trityl-2′-ketouridine (**87a**) in yields of 66 and 62% (*115*). In the same way, oxidation of 2′,5′-di-*O*-tritylcytidine (**84b**) and 3′,5′-di-*O*-tritylcytidine (**86b**), or of their related N^4-acetyl derivatives, gave the corresponding 3′- and 2′-keto derivatives [(**85b**) and (**87b**)] in 40–55% yields (*116*). In the case of the cytidine derivatives, dichloroacetic acid was found to be preferable to pyridinium trifluoroacetate as the proton source. Oxidation of (**84**) and (**86**) could also be achieved using DMSO–P_2O_5 and DMSO–acetic anhydride. (*115,116*). The increased stability of the ditrityl keto nucleosides relative to 3′-ketothymidine derivatives perhaps may be the result of a conformational deformation due to the bulky trityl groups which inhibits the elimination reaction. Alternatively the lability of the ketothymidines may be a consequence of the availability of an activated 2′-proton situated suitably for *trans* elimination of the base. Removal of the trityl groups from (**85**) or (**87**) results in a greatly increased alkaline lability, the pyrimidine base being completely cleaved within seconds at pH 10,

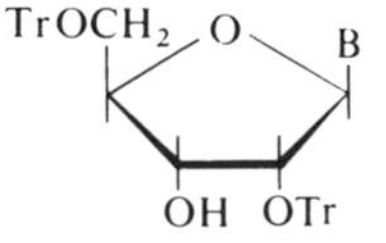

(**84**)(**a**) B = uracil
(**b**) B = cytosine

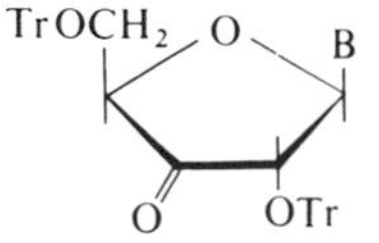

(**85**)(**a**) B = uracil
(**b**) B = cytosine

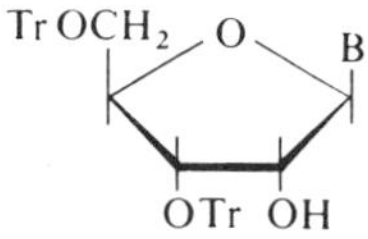

TrOCH$_2$ O B — OTr OH
(**86**)(**a**) B = uracil
(**b**) B = cytosine

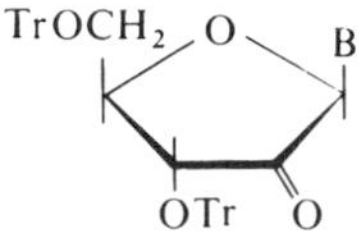

TrOCH$_2$ O B — OTr O
(**87**)(**a**) B = uracil
(**b**) B = cytosine

Finally, oxidation of several polysaccharides has been achieved using DMSO-based oxidants. Bredereck (*117*) has described the oxidation of 6-*O*-tritylcellulose using both DMSO–DCC and DMSO–acetic anhydride and has shown that extensive oxidation occurred rather selectively at C_2 of the glucose residues by isolation of 40–45% of mannose following borohydride reduction and hydrolysis of the polymer. A similar relatively selective oxidation at C_2 of 6-*O*-tritylamylose with DMSO–acetic anhydride has been reported by Wolfrom and Wang (*118*), and Belder et al. (*119*) have described the oxidation of the phenylboronate derivative of a dextran which led to a random attack with introduction of 0.39 carbonyl group per glucose residue. As might be expected, oxidation of starch was more complex, and with DMSO–acetic anhydride there was extensive incorporation of acetyl and thiomethoxymethyl groups, particularly at the primary hydroxyl group (*120*).

3. Applications to Alkaloids

During our studies on the nature of the required acid catalyst for oxidation by the DMSO–DCC method, it was observed that although pyridinium salts of even strong acids were satisfactory, the corresponding triethylammonium salts were not (*20*). This observation suggested that oxidation of hydroxyl groups in molecules containing strongly basic tertiary amine functions would be difficult. **Fortunately, it soon became apparent** (*65,121*) **that if an excess of acid relative to the amine was used, successful oxidation of a number of hydroxy indole alkaloids containing strongly basic tertiary amine functions could be achieved, for example by use of DMSO and DCC in the presence of 1.5 molar equivalents of anhydrous phosphoric acid or, in a few cases, of a small excess of trifluoroacetic acid.** The application of this method to indole alkaloids (*121–125*) is of practical importance since these compounds are notoriously sensitive to most classical methods of oxidation.

In general, the work-up of these oxidation reactions is simplified by the possibility of freeing the product from excess DCC and dicyclohexylurea by extraction of the amine salt into water followed by regeneration of the free base. The yields obtained are generally good, and the method has been used, for example, in the oxidation of yohimbine (**88a → 88b**) (*121*), spegazzinidine dimethyl ether (**89a → 89b**) (*65*), and the hydroxy carbinolamine (**90a → 90b**) (*122*) in yields of 80, 80, and 66%, respectively. The latter example is interesting since the carbinolamine function remains inert and undergoes neither lactam formation nor dehydration (Section II.C.4).

In addition, a number of other indole alkaloids such as α-yohimbine (*121*), methyl reserpate (*121*), 16α-methylyohimban-17α-ol (*121*), 19-dehydroyohimbine (*123*), *N*-methyl-22-hydroxykopsane (*124*), leurosidine (*125*), and several polyfunctional compounds (*126,127*) have been oxidized satisfactorily.

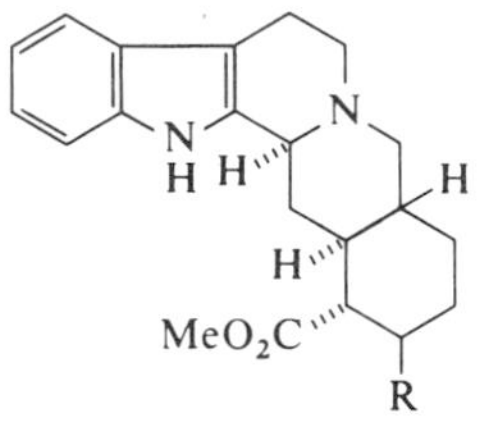

(**88**)(**a**) R = H, α-OH
(**b**) R = O
(**c**) R = H, β-OH
(**d**) R = H, β-OCH_2SCH_3

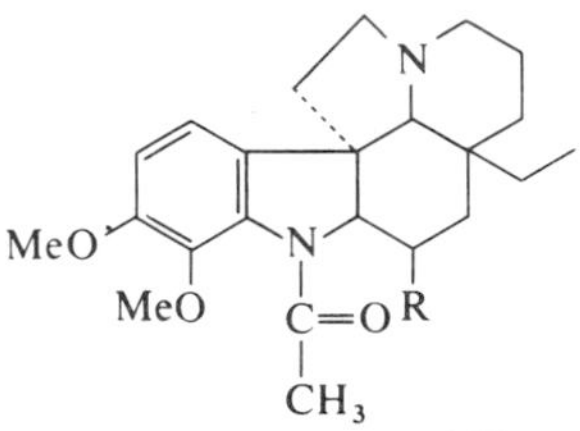

(**89**)(**a**) R = H, α-OH
(**b**) R = O

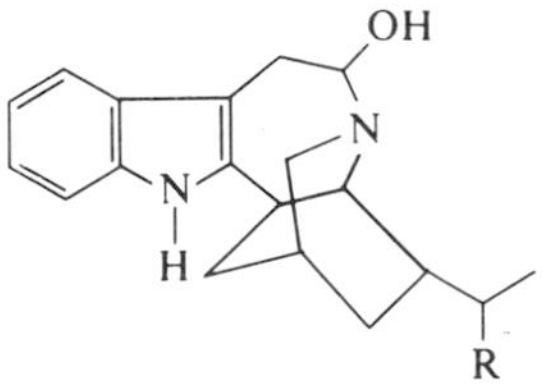

(**90**)(**a**) R = H,OH
(**b**) R = O

Because of their mildness and efficiency, the DMSO–DCC method and the related DMSO–acetic anhydride methods seem admirably suited for use in the alkaloid area.

4. Miscellaneous Examples

During the past few years a variety of hydroxyl groups of different structural types have been subjected to oxidation by the DMSO–DCC method. Once again, one of the most promising applications lies in the oxidation of primary alcohols to exclusively the aldehyde level. Among the simple examples that can be cited are the oxidation of several substituted benzyl alcohols (*65*), various aliphatic primary alcohols (*128–130*) including ^{14}C-propanol (*131*), the terpene alcohol (**91**) (*132*), the hydroxyethylsulfonamide (**92**) (*133*), and the highly unsaturated squalene precursor (**93**) (*134*), all in good to excellent yields. This method offers particular advantages for the preparation of 1-tritiated aldehydes by borotritiide reduction of an existing aldehyde (*74,81,82*) followed by reoxidation and has been applied to the synthesis of the tritiated aldehyde derived from (**93**) (*134*) and of a tritiated lanosterol precursor (*74*).

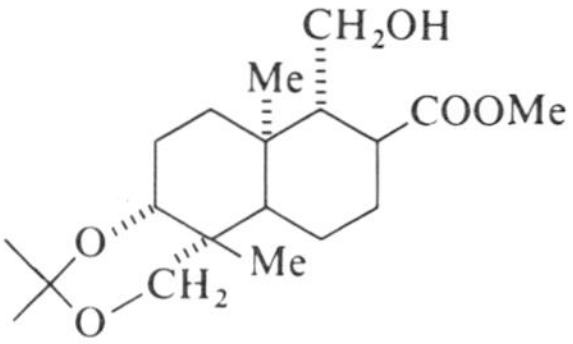

CH_2OH Me COOMe O O CH_2 Me

(**91**)

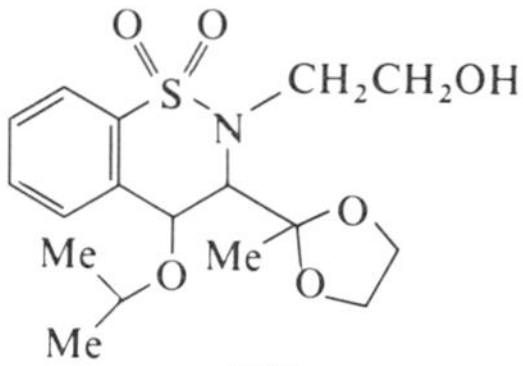

O O S N CH_2CH_2OH Me O Me Me O O

(**92**)

$HOCH_2$

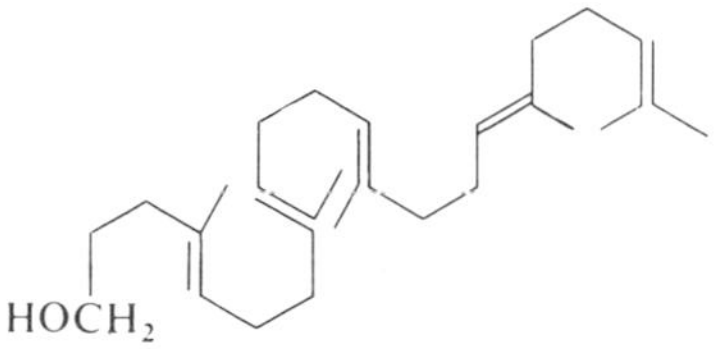

(**93**)

Several alcohols containing both terminal (**94**) and internal (**95**) acetylene or polyacetylene functions have also been oxidized to the aldehyde level without difficulty (*135*).

$$HC{\equiv}C{-}CH_2CH_2CH_2OH \qquad CH_3{-}C{\equiv}C{-}C{\equiv}C{-}CH_2CH_2CH_2OH$$

(**94**) (**95**)

Some steric hindrance is to be found in even certain primary alcohols. Thus Schneider and Meinwald (*136*) have reported that the neopentyl alcohol (**96**) cannot be oxidized using the DMSO–DCC, DMSO–acetic anhydride, Oppenhauer, or chromic oxide–pyridine methods. Modest oxidation, however, could be achieved using chromic oxide in acetic acid under reduced pressure. On the other hand, neopentyl-type alcohols such as Δ^{12}-oleanol (**97**) (*137*) and 1-hydroxymethyl-1,4-methano-1,4-dihydronaphthalene (**98**) (*138*) in which the α-substituents are held in part of a bicyclic ring system, are oxidized to the corresponding aldehydes quite satisfactorily.

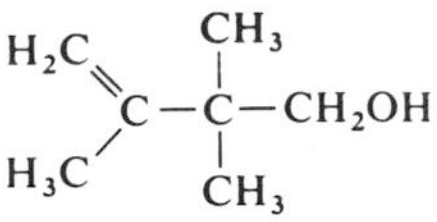

$H_2C{=}C(H_3C){-}C(CH_3)_2{-}CH_2OH$

(96) 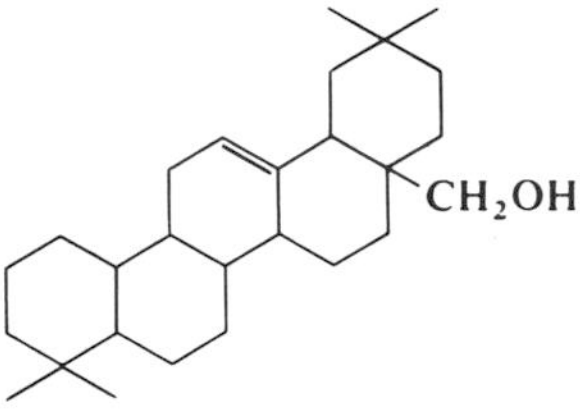 (97) 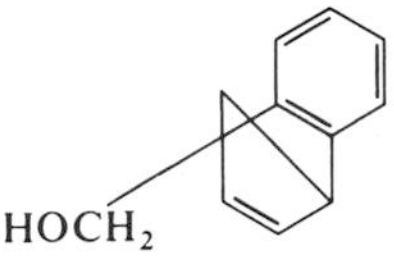 (98)

Among the examples of oxidations of miscellaneous secondary alcohols, the most useful examples are those dealing with alcohols that are generally unstable towards more vigorous reagents. A good example is the oxidation by Paquette and Wise (*139*) of the hydroxy sulfide (**99a**) and some related compounds to the corresponding keto sulfides (**99b**) in quantitative yield using DMSO–DCC and pyridinium trifluoroacetate. The keto sulfide could not be obtained with a variety of other reagents due to the sensitivity of the sulfur atom in the presence of less specific oxidants. An extreme example of the preparation of a highly unstable ketone is found in the synthesis by Dowd and Sachdev (*140*) of 3-methylenecyclobutanone (**100b**) by oxidation of the corresponding alcohol (**100a**). Since the product is very sensitive and undergoes rapid isomerization in the presence of either mild acid or base, the oxidation was carried out using trifluoroacetic acid and quinaldine under vacuum. Under these conditions (**100b**) was distilled directly from the reaction mixture in 20% yield. A complex cyclopropanol (**101a**) has also been oxidized to the corresponding cyclopropanone (**100b**) using DMSO and diisopropylcarbodiimide (*141*).

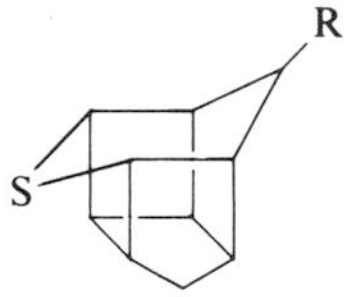

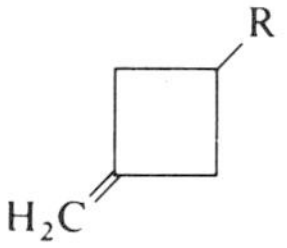

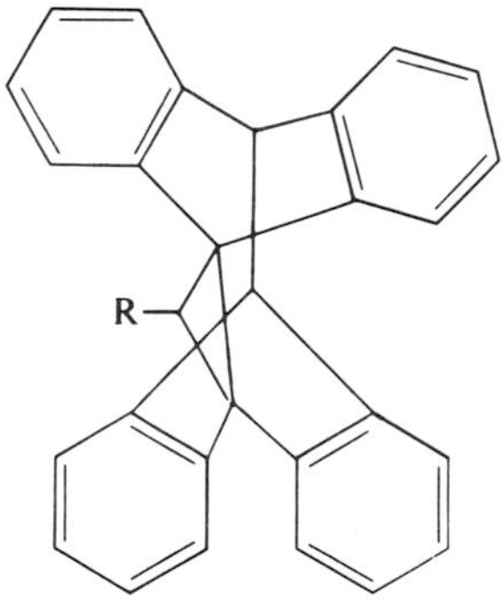

(**99**)(**a**) R = H, OH
(**b**) R = O

(**100**)(**a**) R = H, OH
(**b**) R = O

(**101**)(**a**) R = H, OH
(**b**) R = O

Sensitivity towards side reactions in the presence of conventional oxidants is also a problem when working with compounds containing less common heteroatoms. Thus, attempted oxidation of certain α-silylcarbinols such as (**102a**) with reagents such as chromic acid in acetone leads to extensive cleavage of silicon–carbon bonds. The DMSO–DCC method, however, has proved to be very useful for the synthesis of α-silyl ketones, oxidation of (**102a**) giving the desired (**102b**) in 70% yield (*142*). Extensions to the oxidation of related compounds such as 1-trimethylsilylethanol (**103**) (*143*), bis(triphenylsilyl)methanol (**104**) (*144*), and bis(triphenylgermyl)methanol (**105**) (*144*) were equally successful.

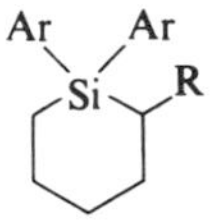

(**102**)(**a**) R = H, OH
(**b**) R = O

$(CH_3)_3Si-CH(OH)CH_3$

$Ar_3SiCH(OH)SiAr_3$ (**104**)

$Ar_3GeCH(OH)GeAr_3$ (**105**)

In addition to the alkaloids and nucleosides referred to earlier there are also examples of the successful oxidation of compounds containing nitrogen heterocycles such as 2-azabicyclo[3,2,1]octan-7-ol (*145*) and highly substituted pyridine derivatives (*146*).

Other than the previously mentioned oxidation of (**82**) to the lactone (**83**) (*110*), only two hemiacetals appear to have been examined with quite different results. Reaction of both epimers of the unsaturated six-membered cyclic hemiacetal (**106a**) with DMSO–DCC led to formation of the same enol lactone (**106b**) (*147*). On the other hand, the five-membered cyclic hemiacetal flindissol (**107**) underwent dehydration to the vinyl ether (**108**), which was isolated in 47% yield (*65*).

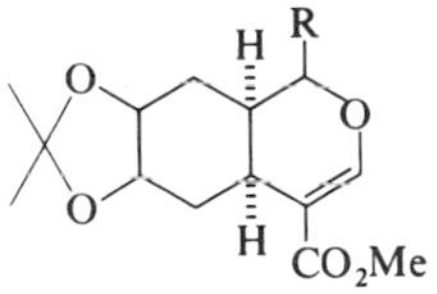

(**106**)(**a**) R = H, OH
(**b**) R = O

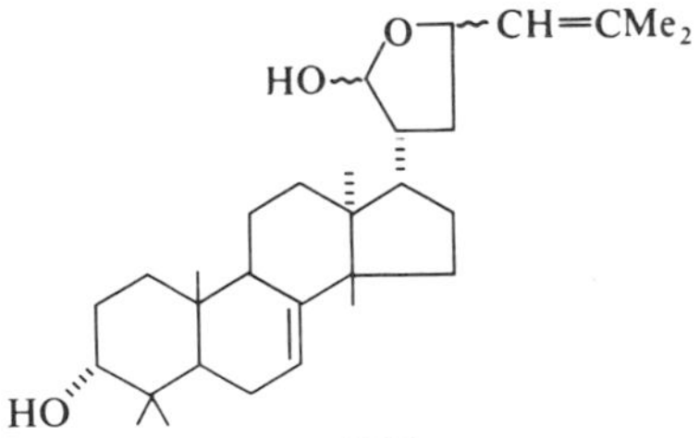

(**107**)

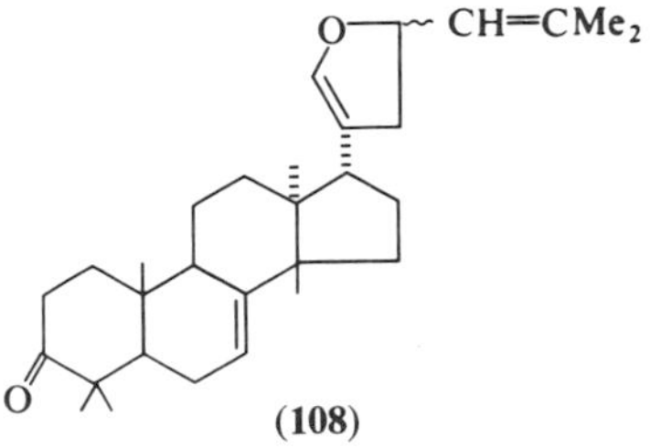

(**108**)

A similar and very rapid dehydration of certain tertiary alcohols also has been observed. Thus, reaction of 17-methyltestosterone (**109**) with DMSO–DCC and pyridinium trifluoroacetate led to almost instantaneous and complete dehydration to a mixture of the Δ^{15} (**110**) and $\Delta^{17(20)}$ (**111**) olefins in a ratio of 3 : 2 (*59*). There was no indication of any D-homo rearrangement accompanying this reaction, and there was an absolute requirement for the presence of DMSO, DCC, and the acid, thus ruling out the possibility of simple acid-catalyzed dehydration.

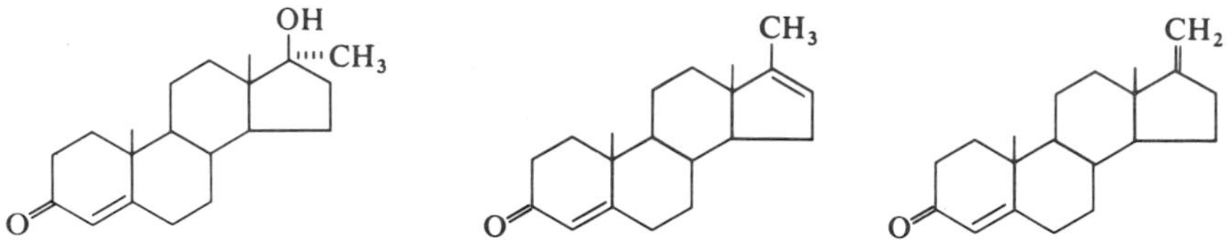

(**109**) (**110**) (**111**)

Such a dehydration would appear to involve an intramolecular proton abstraction mechanism similar to that proposed during oxidation but leading to regeneration of DMSO rather than formation of dimethyl sulfide as in (**112** → **113**).

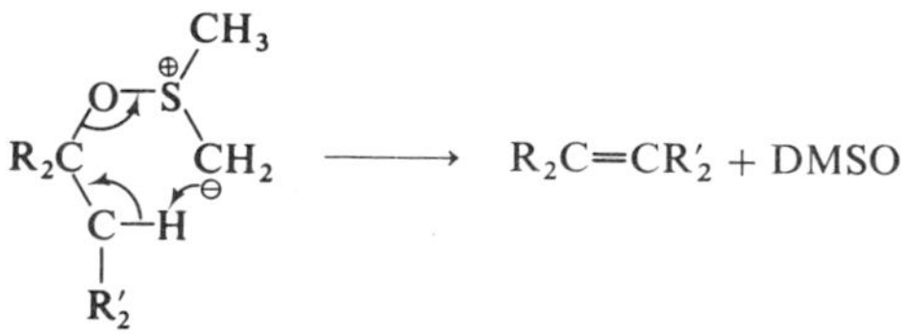

$$R_2C=CR'_2 + DMSO$$

(**112**) (**113**)

Dehydration is not, however, a general reaction of tertiary alcohols, and several examples of selective oxidation of a secondary alcohol in the presence of a tertiary alcohol are known. Treatment of either hydroxy epimer of the furandiol (**114**), for example (*148*), with DMSO–DCC and pyridinium phosphate gave the hydroxy ketone (**115**) in 74% yield, and similar results were obtained with the decalindiol (**116a**), which gave the hydroxy ketone (**116b**) using DMSO in the presence of either DCC and pyridinium trifluoroacetate or acetic anhydride (*149*).

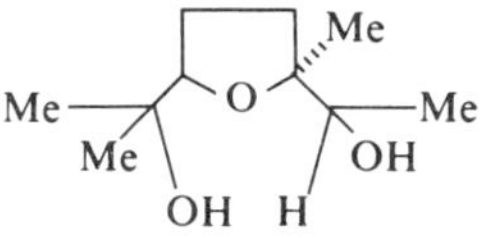

(114)

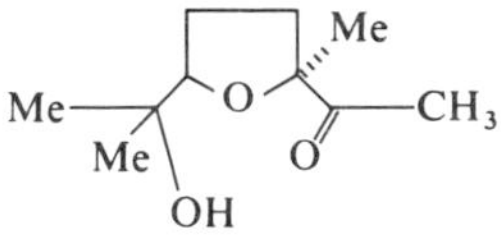

(115)

(116)(a) R = OH
(b) R = O

Allylic alcohols appear to be unique in their reactions and pose some interesting mechanistic questions. We have observed that several simple allylic alcohols such as cinnamyl alcohol (**117**) are readily oxidized to the corresponding conjugated aldehydes upon oxidation with DMSO–DCC and anhydrous phosphoric acid. Quite unlike other alcohols, however, (**117**) is completely inert when pyridinium trifluoroacetate is used as the proton source (*59*). No ready explanation for this unexpected observation is apparent, and further work must be undertaken to more clearly understand it. Also, several steroidal allylic alcohols (**118**) behave anomolously and undergo very rapid dehydration to the heteroanular dienes (**119**) using a variety of proton sources. Once again this is not the consequence of simple acid-catalyzed dehydration, and an absolute requirement for DMSO, DCC, and acid can be demonstrated (*59*). Farnesol (**120**) shows somewhat of a mixture of both effects and undergoes simple oxidation to farnesal using phosphoric acid while being dehydrated to the conjugated diene farnesene using pyridinium trifluoroacetate or dichloroacetic acid (*59*).

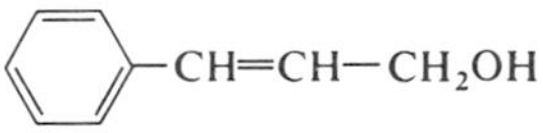

$C_6H_5-CH{=}CH-CH_2OH$

(117)

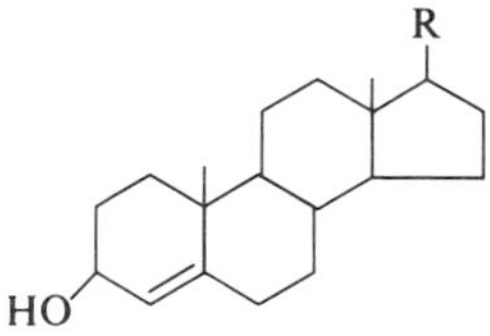

(118)

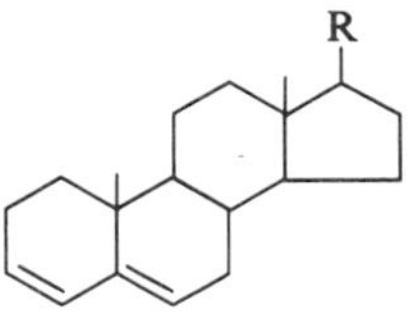

(119)

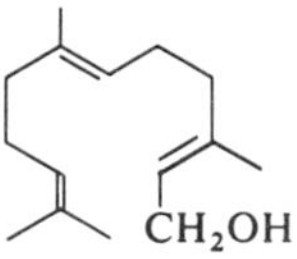

(120)

An apparent exception to this behavior is found in the oxidation in 85% yield of the allylic alcohol (**121a**) to the conjugated ketone (**121b**) using DMSO–DCC and pyridinium phosphate (*150*).

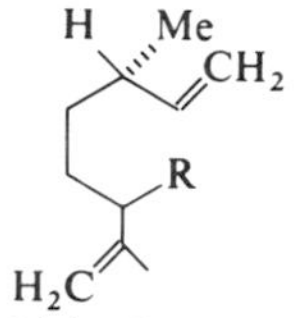

(**121**)(**a**) R = H, OH
(**b**) R = O

Oxidation of free hydroxyl groups by the DMSO–DCC method has been examined as a means of inducing selective bond cleavages in peptides and oligonucleotides. Thus D'Angeli et al. (*151,152*) have shown that oxidation of the secondary hydroxyl group in nitrogen-protected threonine peptides (**122**) leads to β-ketoamides (**123a**) which upon reaction with phenylhydrazine undergo selective peptide bond cleavage via the hydrazone (**123b**) with formation of the pyrazolone (**124**) and an amine.

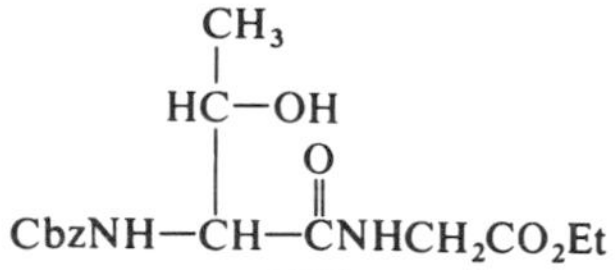

(**122**)

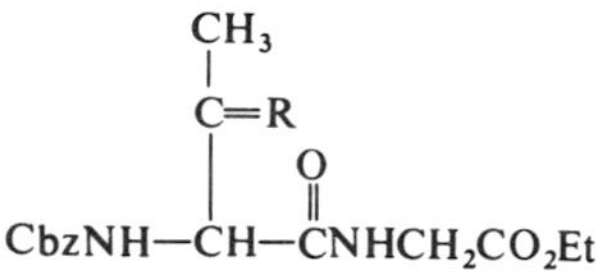

(**123**)(**a**) R = O
(**b**) R = N—NHAr

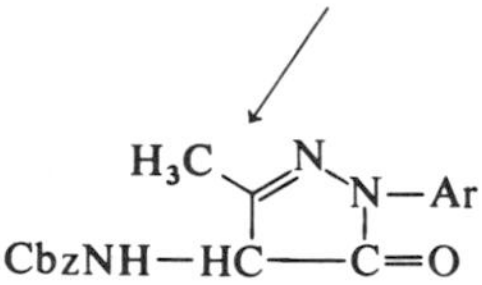

(**124**)

Applied to several simple serine peptides the analogous β-formylamides were formed, but cleavage with phenylhydrazine did not occur. With hydroxylamine, however, cleavage of both the β-keto- and β-formylamides occurred with formation of an *iso*-oxazolone.

Since it is known that oxidation of the 3′-hydroxyl group of thymidine 5′-phosphate (**5**) leads to spontaneous elimination of both thymine and phosphate, similar mild oxidation of deoxyribo-oligonucleotides could provide a method for stepwise degradation of these compounds via the pathway (**125 → 127**). Following enzymatic removal of the 3′-phosphate of (**127**), a second cycle could be initiated. Side reactions between the phosphodiester backbone of the oligonucleotide and the DCC, however, make this process inefficient, and it has been shown (*59,153*) that even at the dinucleotide level the desired cleavage occurs to an extent of only about 50%. The use of DMSO–acetic anhydride for the oxidative step is more promising, and Gabriel et al. (*153*) have demonstrated reasonably selective stepwise degradation of short oligonucleotides.

C B A ... P P P —OH ⟶ [C B A ... P P P =O] ⟶ A + C B ... P P —P

(**125**) (**126**) (**127**)

Finally, it might be mentioned that the reaction of thiols with DMSO–DCC does not lead to the formation of thiocarbonyl compounds but rather to simple disulfide formation. Several examples of disulfide formation in very high yield have been described by Jones and Wigfield (*38*), who also have demonstrated an absolute requirement for DMSO, DCC, and acid in these reactions. It has been suggested that the mechanism of disulfide bond formation involves nucleophilic attack by the sulfhydryl group upon the thiosulfonium intermediate (**128**) arising by attack of the thiol upon the DMSO–DCC adduct (**19**).

$$R'N{=}C(-O-{}^{\oplus}SMe_2)-NHR' + RSH \longrightarrow RS{-}\overset{\oplus}{S}Me_2 \;(\text{attacked by } RSH) \longrightarrow RSSR + Me_2S + H^+$$

(**19**) (**128**)

While this chapter has considered exhaustively the acid-catalyzed reactions of DMSO and DCC with alcohols, it is known that a wide range of other nucleophilic functional groups are also capable of reaction with these reagents. As yet, only the reactions of a variety of phenols (*34,35*) and active methylene compounds (*36*) and oximes (*154*) have been described, but extensive work with many other classes of compounds, leading to some unusual types of products, has been completed and will be described shortly (*33*).

III. DIMETHYL SULFOXIDE–ACETIC ANHYDRIDE OXIDATION OF ALCOHOLS

A. Introduction and Mechanism

Following their work on the application of the DMSO–DCC method to the oxidation of sensitive indole alkaloids (*121*), Albright and Goldman (*30*) developed a very useful alternative reaction using DMSO and acetic anhydride. This reaction is mechanistically closely related to that using DMSO–DCC, and in many cases the two methods are complimentary to one another. In this section, which will not be as detailed as that dealing with the DMSO–DCC method, the various advantages and limitations of both methods will be pointed out.

The mechanism of the DMSO–acetic anhydride oxidation presumably involves an initial reaction between these two components giving the acetoxydimethylsulfonium ion (**129**) in the same way as has been postulated during the Pummerer rearrangement (*54,55*). Rather than undergoing rearrangement, however, (**129**) is attacked by the alcohol with formation of the alkoxysulfonium salt (**130**), which in the presence of acetate ion loses a proton to form the same oxysulfonium ylid (**131**) considered to be the key intermediate in the DMSO–DCC oxidation.

$$(H_3C)_2S{=}O + (CH_3CO)_2O \longrightarrow (H_3C)_2\overset{\oplus}{S}{-}O{-}\overset{O}{\overset{\|}{C}}CH_3 \quad CH_3CO\overset{\ominus}{O}$$

(129)

$$(H_3C)_2\overset{\oplus}{S}{-}OAc + R_2CH{-}OH \longrightarrow (H_3C)_2\overset{\oplus}{S}{-}OCHR_2 + CH_3COOH$$

(129) **(130)**

$$(H_3C)_2\overset{\oplus}{S}{-}O{-}CHR_2 \xrightarrow[-H^{\oplus}]{AcO^{\ominus}} H_3C{-}\overset{\oplus}{S}(\overset{\ominus}{C}H_2){-}O{-}CR_2H \longrightarrow R_2CO + CH_3SCH_3$$

(130) **(131)**

As part of his study on the reactions of alkoxysulfonium salts, Torssell (*58*) reported that the reaction of purified isobutoxydimethylsulfonium tetraphenylborate with DMSO and acetic anhydride fails to give any isobutyraldehyde. This observation does not, however, rule out the intermediacy of alkoxysulfonium ions in the oxidation reaction since it has been pointed out by both Albright and Goldman (*30*) and by Johnson and Phillips (*55*) that under these conditions there is no free acetate anion present which must act as the base to convert the ion (**130**) into the reactive ylid (**131**). It has indeed been demonstrated that the addition of sodium acetate to solutions of alkoxysulfonium fluoroborates in DMSO leads to either the formation of carbonyl compounds or to Pummerer rearrangement in different cases (*55*). The free alkoxysulfonium ion (**130**) is not, in fact, an obligatory intermediate in the oxidation reaction since direct reaction of (**129**) with the alcohol could give a tetracovalent sulfur intermediate (**132**) which is capable of direct collapse via a cyclic process to the sulfonium ylid (**131**) and then to products.

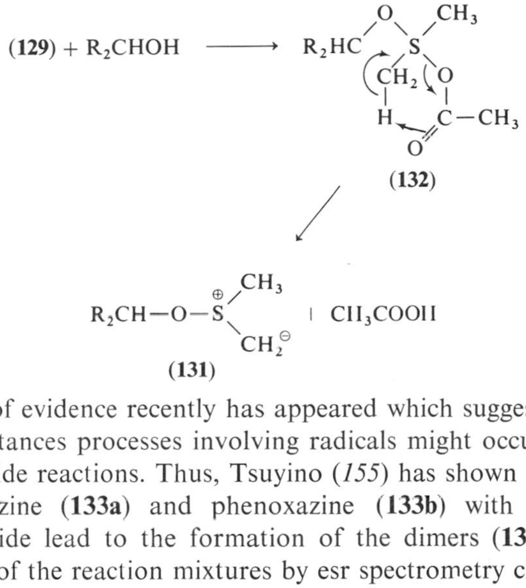

(**129**) + R_2CHOH ⟶ R_2HC–O–S(CH_3)(CH_2–H)(O–C(=O)–CH_3)

(**132**)

$R_2CH{-}O{-}\overset{\oplus}{S}(CH_3)(CH_2^{\ominus})$ + CH_3COOH

(**131**)

One piece of evidence recently has appeared which suggests that under some circumstances processes involving radicals might occur in DMSO–acetic anhydride reactions. Thus, Tsuyino (*155*) has shown that reactions of phenothiazine (**133a**) and phenoxazine (**133b**) with DMSO and acetic anhydride lead to the formation of the dimers (**134**) and (**135**). Examination of the reaction mixtures by esr spectrometry clearly demonstrated the presence of both nitrogen and carbon radicals with structures compatible with the formation of the observed dimers. While these particular cases might be unique, the possibility of radical processes in other sulfoxide reactions should not be overlooked.

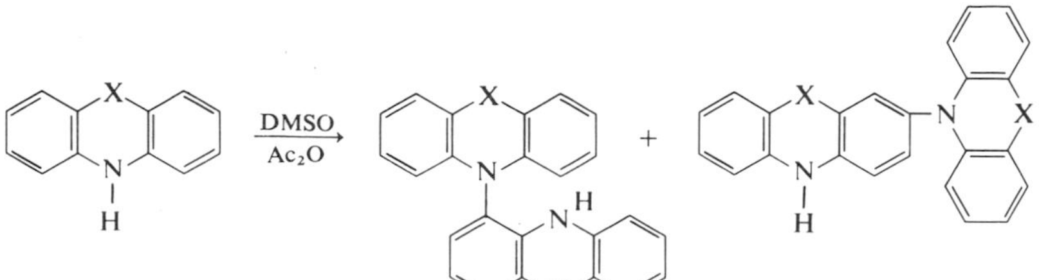

(**133**)(**a**) X = S
(**b**) X = O

$\xrightarrow[Ac_2O]{DMSO}$ (**134**) + (**135**)

The oxidation reaction is usually carried out at room temperature for 12–24 hr using roughly 20 molar equivalents of acetic anhydride and 40 equivalents of DMSO. The reaction mixture is quenched with water and worked up in an appropriate manner. However, the excess of acetic anhydride in certain cases has been reduced to 5 molar equivalents without seriously affecting the oxidation reaction (*30*). As might be expected, other simple sulfoxides such as tetramethylene sulfoxide can be used in place of DMSO, but diphenyl sulfoxide is not a suitable replacement. The nature of the acid anhydride can also be varied, oxidation of yohimbine (**88a**) to yohimbone (**88b**) being readily achieved through reaction with DMSO in the presence of acetic anhydride (84%), benzoic anhydride (82%), phosphorus pentoxide (45%, see later), and polyphosphoric acid (51%). On the other hand, the use of polyphosphoric acid was not satisfactory for oxidation of testosterone (**10**) (*30*), and reactions of several primary aliphatic and benzylic alcohols with DMSO and polyphosphoric acid at 100° have led to the formation of complex mixtures of ethers and acetals and little simple oxidation (*156*). The anhydrides of strong acids such as trifluoroacetic anhydride and *p*-toluensulfonic anhydride are not satisfactory.

B. Scope of the DMSO–Acetic Anhydride Oxidation Reaction

Relative to the use of the DMSO–DCC method, oxidations using DMSO in conjunction with acetic anhydride, phosphorus pentoxide (*31*), or pyridine–sulfur trioxide (*32*) offer the distinct advantage of producing mainly water-soluble by-products. While destruction of excess DCC usually can be efficiently achieved using treatment with oxalic acid as previously described (Section II,A), removal of the last traces of the resulting dicyclohexylurea is sometimes difficult and requires chromatography. During the use of DMSO–acetic anhydride, however, the by-product is predominantly acetic acid, which can be removed by aqueous extraction together with the DMSO. When dealing with alcohols which are appreciably soluble in water, removal of most of the reaction by-products can be achieved by lyophilization without the necessity of extraction (*110*). While this feature of the DMSO–acetic anhydride reaction is very attractive and convenient, the method is not applicable to the same wide range of alcohols as its DCC counterpart. Thus, **the acetic anhydride reaction is admirably suited to the oxidation of rather hindered alcohols,** but its application to relatively unhindered cases leads to extensive formation of acetate esters and thiomethoxymethyl ethers.

The choice of reagent in each potential application must therefore be made taking into account both the efficiency of the oxidation reaction per se and the ease of work-up of the reaction mixture.

Successful oxidations of many types of hydroxyl functions have been achieved using the DMSO–acetic anhydride method, and some specific comments relating to these examples will be found in the following sections.

1. Applications to Steroids and Alkaloids

As was found to be the case with DMSO–DCC oxidations, the relatively rigid conformations of the steroid and alkaloid skeletons have provided considerable information as to the structural requirements for DMSO–acetic anhydride reactions. Thus, Albright and Goldman (*30*) have shown that while oxidation of the axial 17-hydroxyl group of yohimbine (**88a**) gives yohimbone (**88b**) in 84% yield and with only traces of by-products, the epimeric, equatorial 17-hydroxyl of β-yohimbine (**88c**) gives an equal mixture of the desired ketone and the thiomethoxymethyl ether (**88d**). Similar formation of substantial amounts of thiomethoxymethyl ethers accompanied the oxidation of the equatorial hydroxyl groups of the other indole alkaloids, α-yohimbine, and methyl reserpate (*30*). However, the method has been very useful for the oxidation of otherwise sensitive indole alkaloids such as ajmaline (*30*) (in which the carbinolamine function remains inert), fruticosin (*157*), and conoflorinol-A and -B (*158*), and also has been applied to compounds in the benzoquinoline (*159*) and isoquinuclidine (*128*) series. Successful oxidation of the allylic hydroxyl groups of galanthamine giving narwedin also has been reported (*30*).

An extreme example of the effect of steric environment was found using 11-hydroxysteroids. Thus, while oxidation of the 11β-hydroxyl group of hydrocortisone-21-acetate (**36d**) was slow and required 3 days at room temperature, the corresponding 11-ketone was obtained in 53% yield. In contrast, the equatorial hydroxyl of 11α-hydroxyprogesterone (**36a**) gave only 13% of the 11-ketone (**36c**), and the major product was the thiomethoxymethyl ether (**36e**) (*30*). These results differ markedly from those obtained using DMSO–DCC where the equatorial alcohol (**36a**) was readily oxidized and its axial counterparts (**36b, d**) were largely inert. As mentioned earlier, these results seem best explained by the greater steric bulk of the DMSO–DCC adduct (**19**) relative to that of the acetoxy-dimethylsulfonium ion (**129**), which will tend to somewhat impede the DMSO–DCC oxidation of highly hindered axial alcohols. On the other hand, DMSO–acetic anhydride mixtures are clearly a much richer source of the methyl methylenesulfonium ion (**32**), and this species, as well as acetic anhydride itself, can compete favorably with (**129**) for attack on unhindered alcohols leading to the thioethers and acetates, respectively. Some evidence has been presented that direct aklylation of an alcohol by (**32**) may not be the sole route by which thiomethoxymethyl ethers are formed. Thus, Ifzal and Wilson (*160*) have shown that reaction of either cholesterol (**136**) or *i*-cholesterol (**138**) with DMSO and acetic anhydride at 100° leads to the formation of the same thiomethoxymethyl ether (**137**) in 70% yield together with several other minor common products. These results have been interpreted in terms of an equilibration of the initial alkoxysulfonium ions (**139** and **141**) by way of the carbonium ion (**140**) followed by rearrangement of (**139**) to the observed ether (**137**). While the latter rearrangement might occur at 100°, the arguments presented by Albright and Goldman (*30*), which have been mentioned earlier, would appear to make this unlikely at room temperature, and the results of comparable experiments at lower temperature would be of interest.

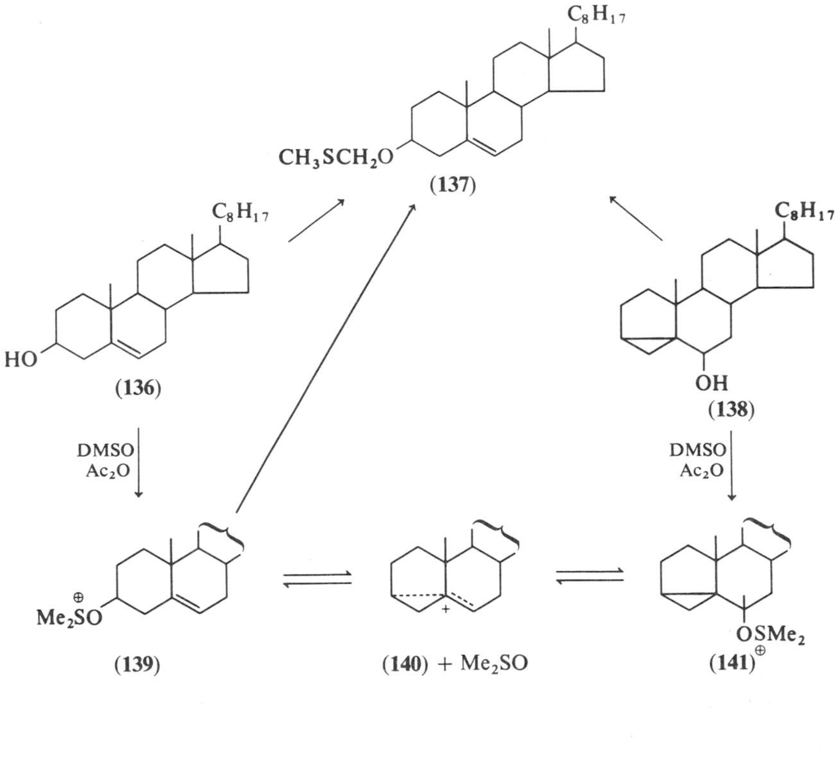

C_8H_{17}
CH_3SCH_2O
(**137**)
C_8H_{17}
C_8H_{17}
HO
(**136**)
OH
(**138**)
DMSO
Ac_2O
DMSO
Ac_2O
$Me_2\overset{\oplus}{S}O$
(**139**)
(**140**) + Me_2SO
$O\overset{\oplus}{S}Me_2$
(**141**)

Certainly the nature of the products in such reactions is a function of temperature. Thus, Ifzal and Wilson (*160*) have reported the formation of only the simple acetate upon reaction of cholestanol (**142a**) with DMSO and acetic anhydride at 100°, but comparable experiments at room temperature (*60*) gave 21% ketone (**142b**), 39% thiomethoxymethyl ether (**142c**), and 33% acetate (**142d**). In comparison, this unhindered alcohol gave the ketone in 80% yield with the DMSO–DCC method, and even under the most unfavorable conditions, using phosphoric acid as the proton source, only 8% of the ether (**142c**) was isolated (*65*). Only occasional other oxidations have been reported at elevated temperatures, two examples being the conversion, in unspecified yield, of a porphyrin containing the partial structure (**143**) into the corresponding ketone (*161*) and the successful oxidation of the *N*-acetyl derivatives of (**84b**) and (**86b**) (*116*). In some cases, the formation of acetates becomes, for unexplained reasons, the principal reaction as, for example, in attempts to oxidize quinine, cinchonine, and related compounds (*162*). Successful oxidations of the secondary hydroxyl groups in testosterone (*30*) and in pregnane-17,20β-diols (*163*) have been reported.

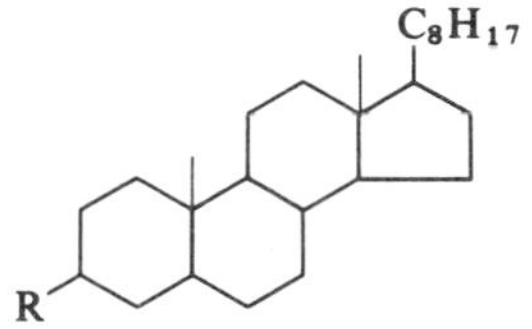

C_8H_{17}

R

N OH

(**142**)(**a**) R = β-OH, H
(**b**) R = O
(**c**) R = β-OCH_2SCH_3, H
(**d**) R = β-OA_c, H

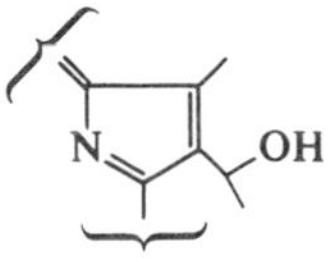

(**143**)

2. Applications to Carbohydrates

The DMSO–acetic anhydride method has found its most frequent and valuable applications to date in the oxidation of isolated hydroxyl functions in carbohydrates. The oxidation of hydroxyl groups at most positions in sugar molecules can be achieved with notable success. The conversion of the hemiacetal function into a lactone appears to be particularly facile and has been achieved, for example, in 95, 100 and 84% yields in the cases of 2,3,5-tri-*O*-benzyl-D-arabinofuranose (**144**) (*94*), 2,3;5,6-di-*O*-isopropylidene-D-mannofuranose (**145**) (*110*), and 2,3,4,6-tetra-*O*-benzyl-D-glucopyranose (*164*), respectively.

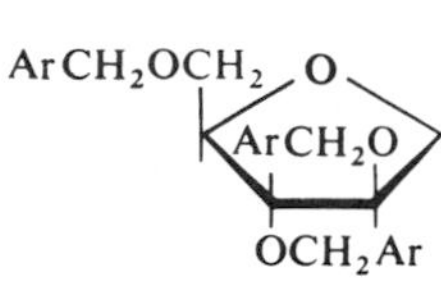

(144) 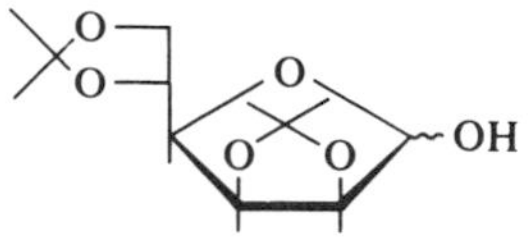 (145)

The development of mild and efficient methods of oxidation based upon DMSO has stimulated much interest in the synthesis of unusual sugars by oxidation of an isolated hydroxyl group followed by sterospecific reduction to the more hindered, and frequently inverted, alcohol. Towards this end, there have been many reported oxidations of the secondary hydroxyl groups in sugar derivatives using the DMSO–acetic anhydride method. Most of these proceed in satisfactory yield and without unexpected side reactions other than the formation of minor amounts of the thiomethoxymethyl and acetyl derivatives of the alcohol.

Typical inversion procedures have been used to prepare the tallose, allosamine, and gulose derivatives [(**146c**), (**147c**), and (**148c**)] from the corresponding galactose (**146a**) (*165,166*), glucosamine (**147a**) (*167*), and galactose (**148a**) (*168*) epimers via oxidation to the keto sugars [(**146b**), (**147b**), and (**148b**)] in yields of 90, 59, and 44% using DMSO and acetic anhydride.

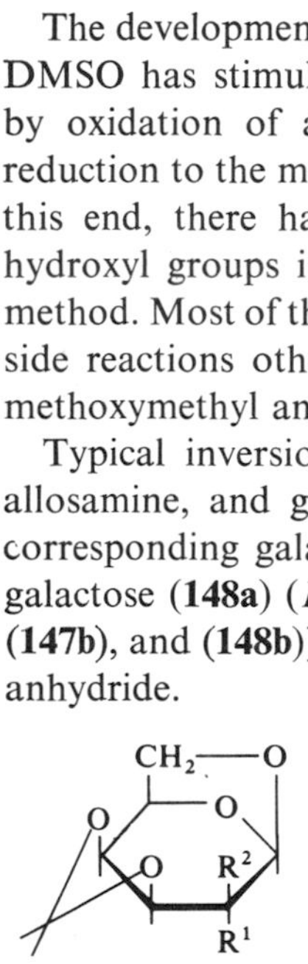

(**146**)(**a**) $R^1 = OH, R^2 = H$
(**b**) $R^1, R^2 = O$
(**c**) $R^1 = H, R^2 = OH$

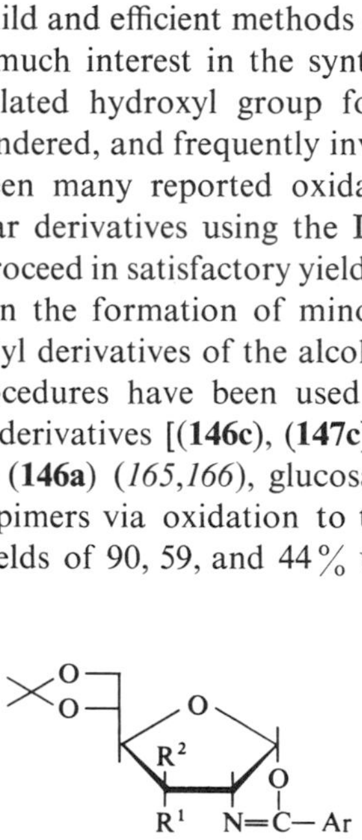

(**147**)(**a**) $R^1 = H, R^2 = OH$
(**b**) $R^1, R^2 = O$
(**c**) $R^1 = OH, R^2 = H$

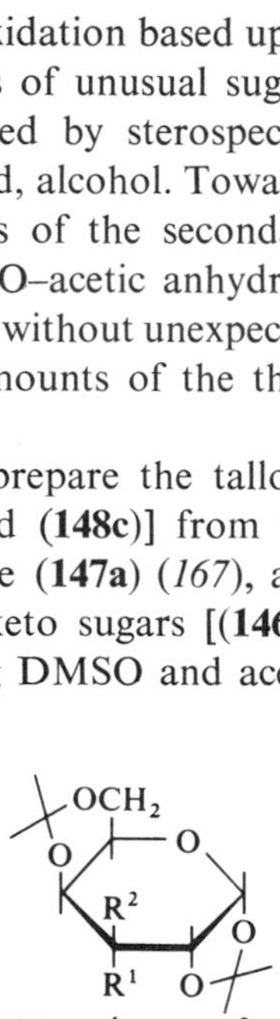

(**148**)(**a**) $R^1 = H, R^2 = OH$
(**b**) $R^1, R^2 = O$
(**c**) $R^1 = OH, R^2 = H$

In addition to (**146a**) several other suitably protected galactopyranoses have been oxidized to 2-keto derivatives (*169,170*) as have the arabinitol (**61**) (*94*) and the ditrityluridines (**84a**) and (**86a**) (*115*). Oxidation of the cytidine derivatives (**84b**) and (**86b**) was accompanied by acetylation of the 4-amino functions (*116*). 3-Keto derivatives of suitably protected xylofuranose (*171,172*) xylopyranose (*173*), glucofuranose (*102,104,174,175*), glucopyranose (*176,177*), 2-deoxyglucopyranose (*169,170*), galactopyranose (*168*), allopyranose (*178*), and fructopyranose (*101–103*) sugars have been reported. Preparations of 4-keto sugars are less frequent but have been accomplished in the mannopyranose (*110,165*), galactopyranose (*179*), and *altro*-heptulopyranose (*180*) series. Similarly 5-keto compounds have been obtained from suitable glucofuranose (*113,181*), and galactofuranose (*182*) precursors as well as from open-chain derivatives of sorbitol (*94*) and gluconamide (*113*). As was mentioned previously, oxidation of (**78**) was accompanied by epimerization of the axial azido function (*107*).

A useful modification of the oxidation-reduction sequence uses reduction of the keto sugars with sodium borotritiide, a procedure that has been used successfully for the preparation of 3-^{3}H-ribose (*174*) and 4-^{3}H-glucose (*179*).

The ready availability of keto sugars has also led to new syntheses of branched chain and amino sugars. For examples Rosenthal (*178*) has oxidized methyl 4,6-benzylidene-2-deoxy-α-D-allopyranoside (**149c**) in 90% yield and converted the product (**149b**) into a carbomethoxymethylidene derivative (**149c**) via reaction with a phosphonate carbanion. Amino sugars such as 3-amino-3-deoxyribose (**150b**) have been prepared by reduction of the oxime derived from keto sugars such as (**150**) (*171*).

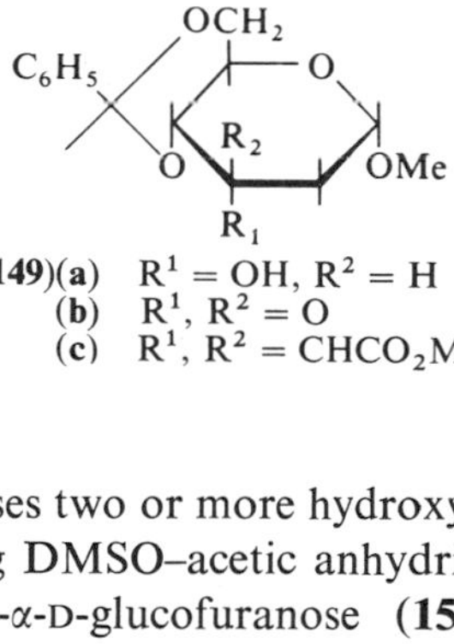

OCH$_2$ C$_6$H$_5$ O R$_2$ O OMe R$_1$

(**149**)(**a**) R^1 = OH, R^2 = H
(**b**) R^1, R^2 = O
(**c**) R^1, R^2 = CHCO$_2$Me

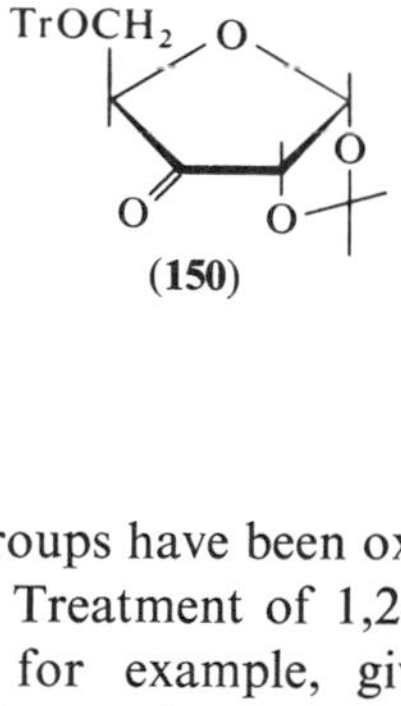

TrOCH$_2$ O O O O

(**150**)

In several cases two or more hydroxyl groups have been oxidized simultaneously using DMSO–acetic anhydride. Treatment of 1,2-*O*-isopropylidene-6-*O*-tosyl-α-D-glucofuranose (**151**), for example, gives the 3,5-diketone in a single step (*182*), and, while the product was not isolated but rather directly reduced, oxidation of 2,7-anhydro-4,5-*O*-isopropylidene-β-D-*altro*-heptulopyranose gave presumably the corresponding keto aldehyde (*183*). An extreme example is the conversion of several free inositols [e.g., (**152**)] into pentaacetoxybenzene (**153**) in yields of 54–60% upon brief reaction with DMSO, acetic anhydride, and pyridine at 65° (*184*). This reaction appears to involve the intermediate formation of a diketoinositol which undergoes enolization, dehydration, and acetylation. Other substituted inositols such as 1,3,4,5,6-penta-*O*-methyl-(−)-inositol undergo simple oxidation (*185*).

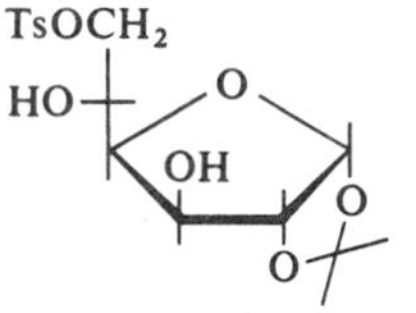

TsOCH$_2$ HO OH O O O **(151)**

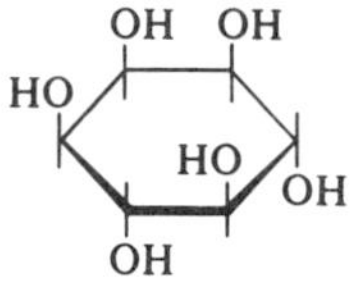

OH OH HO HO OH OH **(152)**

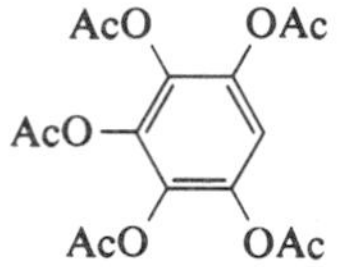

AcO OAc AcO AcO OAc **(153)**

In contrast to the very successful oxidation of a wide range of carbohydrate secondary hydroxyls, the DMSO–acetic anhydride method seems to be generally unsuitable for the preparation of aldehydes from primary hydroxyl groups. Thus, attempted oxidations of 1,2;3,4-di-*O*-isopropylidene-α,D-galactopyranose (**53a**) (*186*) and of methyl 2,3,4-tri-*O*-acetyl-α,D-glucopyranoside (*89*) have led primarily to the corresponding thiomethoxymethyl ethers and acetate esters with no more than traces of the desired aldehydes being formed. **Since both aldehydes are readily prepared using the DMSO–DCC method** (*83,88,89*), **the latter method is clearly to be preferred for use with primary alcohols.** In the case of certain secondary hydroxyl groups, however, the acetic anhydride method appears to provide higher yields of ketones (*110*). In all cases, it is useful to recall that since homogeneous reaction mixtures are generally obtained using the DMSO–acetic anhydride method, it is frequently possible to kinetically follow the course of the oxidation by polarimetry (*164*).

Oxidation of a few polysaccharides has also been examined, 2-keto derivatives being the predominant products from starch (*120*), 6-tritylcellulose (*117*), and 6-tritylamylose (*118*), while dextran (*119*) and dextran-2,4-*O*-phenylboronate (*119*) gave, at least partially, the 3-ketones.

3. Miscellaneous DMSO–Acetic Anhydride Oxidations

Although it has been mentioned several times that primary hydroxyl groups are generally not oxidized in a satisfactory way using DMSO–acetic anhydride, this is not always the case. Several relatively hindered primary alcohols such as (**154**) (*187*) and (**155**) (*188*) have been oxidized in very high yields to the corresponding aldehydes. In the case of (**154**), the tertiary alcohol remains unchanged, and similar results have been obtained with other compounds (*163,189*) such as (**116a**) (*149*). In the case of the triol (**156**), the diketoaldehyde was obtained in a single step although the yield was only 11% (*190*).

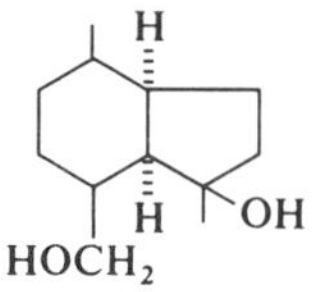

(154) 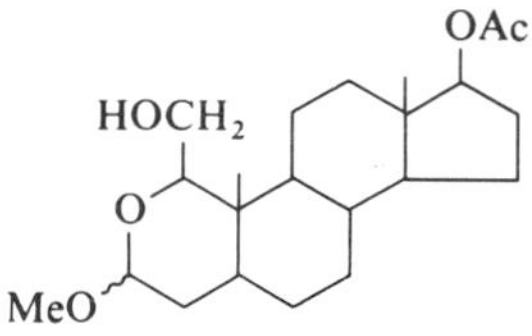 (155) 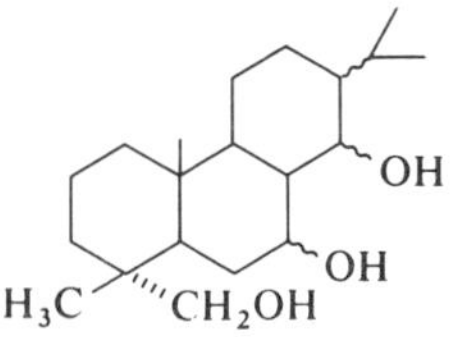 (156)

Oxidation of a neopentyl type of alcohol in 2-(2-hydroxy-1,1-dimethylethyl)-indole to the corresponding aldehyde also has been achieved (*191*).

Some interesting examples of the effects of steric hindrance upon the course of the reaction are found in a study by Bhatia et al. on the oxidation of various flavanols (*192*). These workers have shown that while flavan-4-ols (**157a**, **157b**) are readily oxidized to the corresponding flavanones, and 1-epi-catechin tetramethyl ether, (**158**) is also rather slowly converted into the previously uncertain 3-keto derivative, the presence of a methoxyl group at C-5, as in (**159**), almost completely blocks all oxidation at C-4. A C-5 methoxyl does not, however, impede oxidation of a C-3 hydroxyl, and both leucocyanadin tetramethyl ether (**160a**) and taxifolin (**160b**) are readily converted into the corresponding C-3 ketones [(**161a**) and (**161b**)]. Oxidation of these 3-hydroxyflavans does not occur using chromic oxide in pyridine.

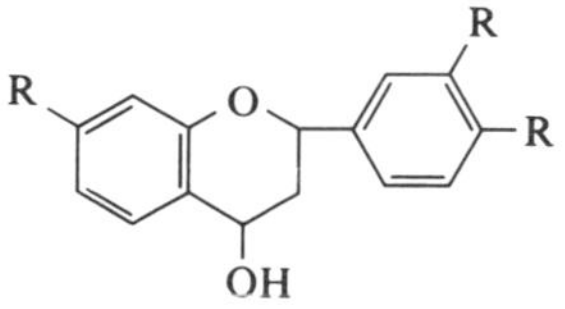

(**157**)(**a**) R = H
(**b**) R = OMe

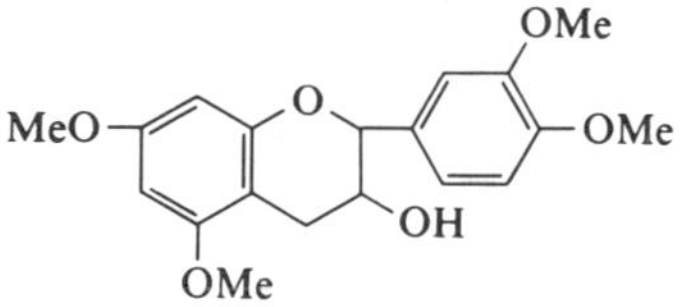

(**158**)

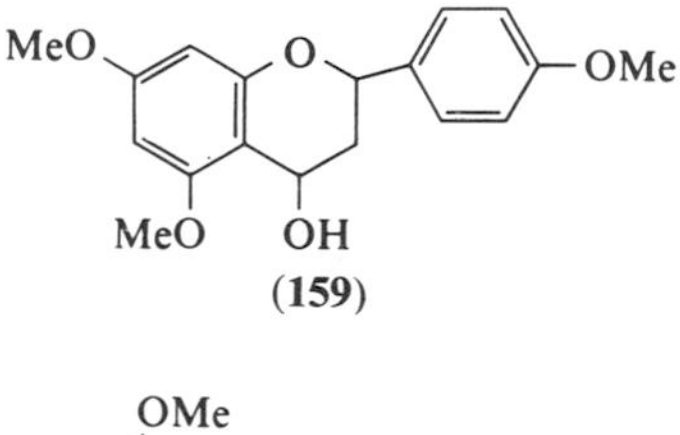

(**159**)

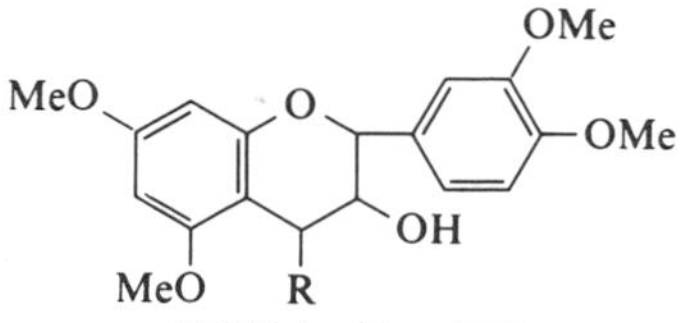

(**160**)(**a**) R = OH
(**b**) R = O

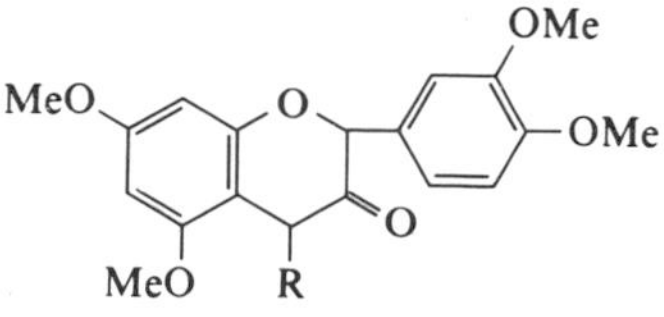

(**161**)(**a**) R = OH
(**b**) R = O

Few attempts have been made to oxidize vicinal diols using sulfoxide based reagents. However, Newman (*193*) has shown that the dihydrobenzanthracene-diol (**162**) is readily converted by DMSO–acetic anhydride into the orthoquinone (**163**) in 47% yield; other methods of oxidation are unsatisfactory. On the other hand, a simple aliphatic diol, butane-2,3-diol, was not satisfactorily oxidized. Also, while the tetramethylhydroquinone (**164a**) was quantitatively converted to duroquinone (**165**), hydroquinone itself (**164b**) gave a complex array of products. With the DMSO–DCC method, however, the oxidation of (**164b**) was quantitative (*34*).

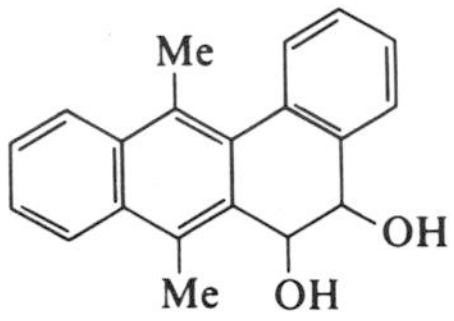

(**162**) 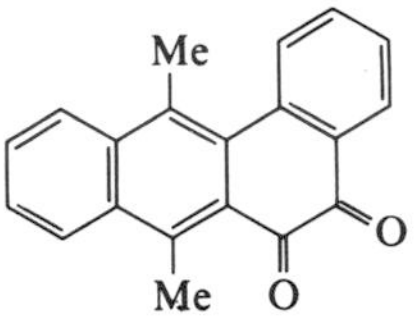 (**163**) 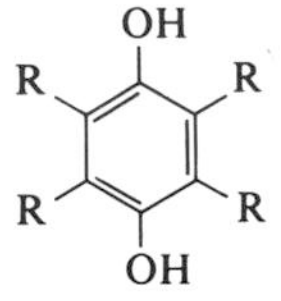 (**164**)(**a**) R = CH_3 (**165**)

(**b**) R = H

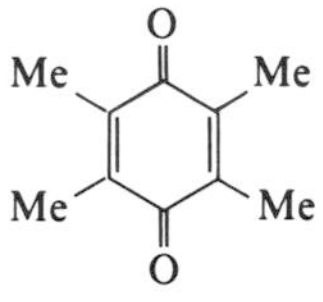

Along similar lines, it has been shown (*194*) that while aromatic acyloins such as benzoin (**166a**), anisoin (**166b**), and furoin are efficiently converted into diketones, their aliphatic counterparts such as 2-hydroxycyclohexanone give very low yields. An exception to this rule is found in the tricyclic four-membered cyclic acyloin (**167**), which has been oxidized to the corresponding cyclobutanedione in 75% yield (*195*).

ArC(=O)—CH(OH)—Ar

(**166**)(**a**) Ar = C_6H_5 (**167**)

(**b**) Ar = $C_6H_4OCH_3$

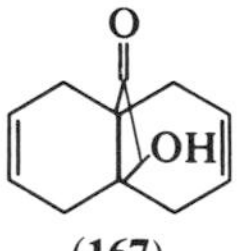

A somewhat exotic example of the utility of DMSO and acetic anhydride is in the conversion of polyporic acid (**168**) into pulvinic acid dilactone (**171**) in 95% yield (*196*). While not strictly oxidative, this intriguing rearrangement probably involves the initial formation of the sulfonium salt (**169**) in much the same way that other 1,3-diketones react with DMSO and DCC or acetic anhydride (*36*). Subsequent attack by acetate then leads to the formation of the ketene (**170**), which, by two successive intramolecular additions of enolic hydroxyl groups to the ketene and to the mixed anhydride, gives the final dilactone.

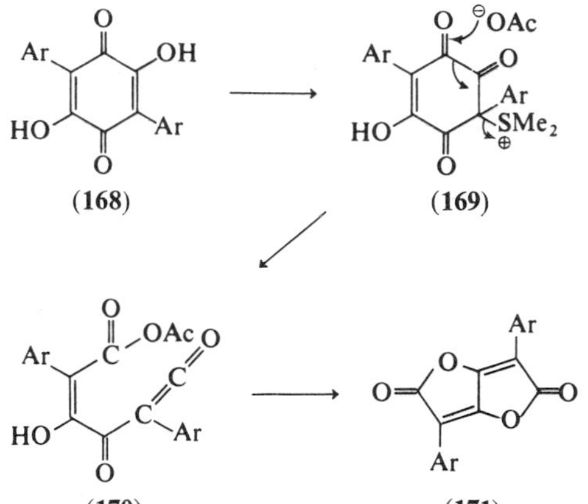

(168) (169)

(170) (171)

Several monohydroxyquinones (**172**) also react with efficient formation of γ-arylidenebutenolides (**174**), and once again, a ketene intermediate (**173**) seems likely (*196*).

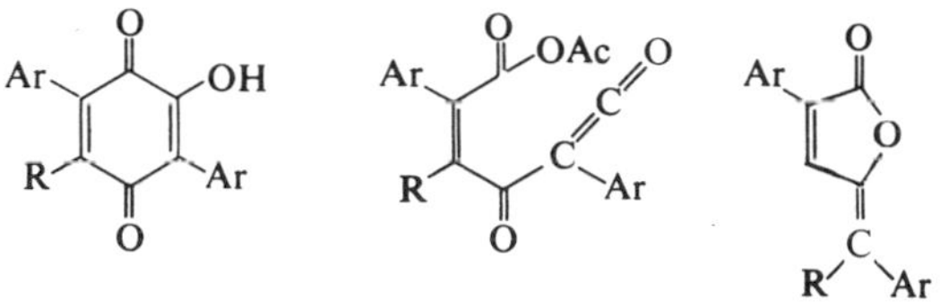

(**172**) R = H, Cl, CMe (**173**) (**174**)

A number of other isolated examples may be cited (*197*), many of those dealing with oxidation of suitable derivatives of natural products such as phaophorbid-a (*161*), polyerythrin-A (*198*), fusicoccin (*199*), and aranotin (*200*).

As was the case with DMSO–DCC reactions, a variety of functional groups other than alcohols have been found to react with DMSO activated by acetic anhydride. In particular, the reactions of phenols (*201–203*) and active methylene compounds (*204,205*) might be mentioned as well as a number of nitrogenous functions (*33*). Since, however, many of these reactions are not oxidative in nature, they will not be considered further at this time.

IV. OXIDATIONS WITH DMSO AND INORGANIC ACID ANHYDRIDES

A. Use of DMSO–Phosphorus Pentoxide

During the course of a general study on the use of phosphorus pentoxide as a dehydrating agent for the synthesis of polysaccharides, Onodera et al. (*206*) observed that treatment of 2,3,4,6-tetra-*O*-acetyl-*β*-D-glucopyranose (**175**) with phosphorus pentoxide in DMSO at 65° led to an unusual conversion to methyl 2,3,4,6-tetra-*O*-acetyl-D-gluconate (**176**). Whether the methyl ester originated from DMSO or from methanol used during the work-up has not yet been ascertained. The oxidative nature of the reaction was not, however, in doubt, and Onodera et al. (*31*) have carefully developed the use of DMSO and phosphorus pentoxide as an efficient oxidant for secondary alcohols, particularly in the field of carbohydrates.

Almost certainly the mechanism of the reaction is closely related to that using DMSO and acetic anhydride, with formation of an initial activated sulfoxonium derivative with the structure (**177**). Subsequent attack by the alcohol will then give the same alkoxysulfonium salt (**130**) or alkoxysulfonium ylid (**131**) intermediates previously discussed. As in the case of DMSO–acetic anhydride reactions, this method offers the distinct advantage that all of the by-products are soluble in water and can readily be removed together with the DMSO by aqueous extraction.

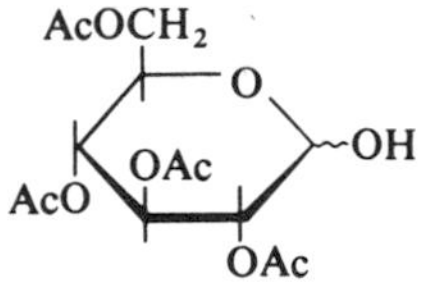

(**175**) 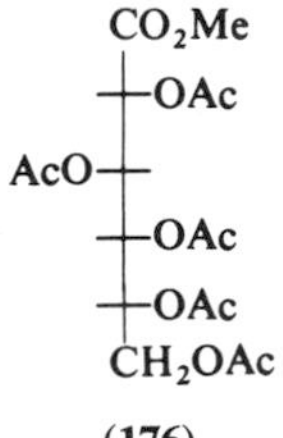 (**176**) 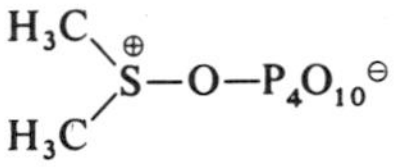 (**177**)

Onodera et al. (*31*) have concluded that **optimal reaction conditions use 1.0–1.5 molar equivalents of phosphorus pentoxide (as P_4O_{10}) and 3–4 molar equivalents of DMSO in dimethylformamide at 65–70° for 1.5–2.0 hr.** If DMSO itself is used as the solvent it is better to perform the oxidation at room temperature for 15–20 hr; otherwise the yields are reduced. With the exception of a brief reference to the failure of the method to cleanly oxidize ajmaline (*30*), all recorded examples have been with carbohydrates. In fact, the majority of the recorded examples have dealt with the oxidation of the 3-hydroxyl group of variously substituted sugars such as 1,2;5,6-di-*O*-isopropylidene-α-D-glucofuranose (**66**) (*31,207*) and 4,6-*O*-ethylidene-1,2-*O*-isopropylidene-α-D-galactopyranose (**148a**) (*168*) in yields of 65 and 63%, respectively. Most conventional protecting groups appear to be stable under the reaction conditions. The yields tend to be good, and there seems to be little difference in the isolated yields during oxidation of several pairs of epimeric alcohols. Thus, oxidation of methyl 4,6-*O*-benzylidene-2-*O*-tosyl-α-D-glucopyranoside (**71**) and of its allofuranose epimer led to isolation of the same 3-keto sugar in yields of 92 and 80%, respectively (*31*). Similar, although lower, yields were also obtained during oxidation of 1,2-*O*-isopropylidene-5-*O*-tosyl-α-D-ribofuranoside (**178a**) and its xylo epimer (**178b**) (*31*) and of methyl 2-acetamido-4,6-*O*-benzylidene-2-deoxy-α-D-glucopyranoside and its α-D-allo epimer (*31*).

The reaction conditions used have almost always been those specified by Onodera et al. (*31*), an exception being the addition of pyridine during the oxidation, in 81% yield, of methyl 6-deoxy-2,3-*O*-isopropylidene-α-D-mannopyranoside (**179**) by Stevens et al. (*208*). The latter modification might prove useful during the oxidation of compounds containing acid-labile functions. By comparison, even after five consecutive treatments of (**179**) with chromic oxide in pyridine, only a 30% yield of the corresponding ketone could be obtained (*208*).

It is interesting to note that oxidation of (**178b**), or of the analogous 5-*O*-diphenylphosphate ester, could be achieved without apparent elimination of the sulfonate or phosphate esters, although the yields were only 28 and 35% (*31*). This method has also been used successfully for the oxidation of 1,3,4,5,6-penta-*O*-benzyl-*myo*-inositol (*209*), 5-*O*-benzyl-1,2-*O*-isopropylidene-α-D-xylofuranose (*210*), and 1,2-*O*-isopropylidene-α-D-glucofuranurono-6,3-lactone (*31*); the latter compound gave the α-keto lactone in 40% yield.

No examples of the oxidation of primary alcohols have appeared as yet, but it might be expected that the formation of thiomethoxymethyl ethers could become troublesome as in the case of DMSO–acetic anhydride. Certainly, in our hands, the phosphorus pentoxide method has not proved satisfactory for the preparation of 2′,3′-protected nucleoside 5′-aldehydes (**44**), although 2′- and 3′-ketonucleosides (**85, 87**) can be readily prepared from suitable intermediates (**84, 86**) (*115*).

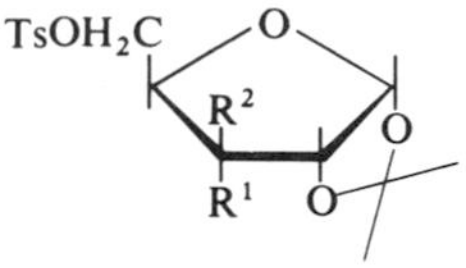

(178)(a) $R^1 = OH, R^2 = H$
(b) $R^1 = H, R^2 = OH$

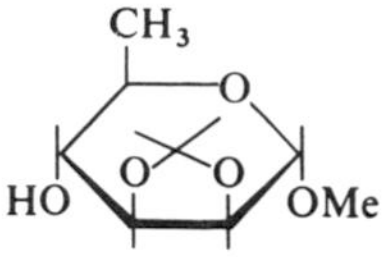

(179)

B. Use of DMSO–Sulfur Trioxide

Although Onodera et al. (*31*) have mentioned briefly that sulfur trioxide and DMSO do not provide an oxidizing media, Parikh and Doering (*32*) have shown that the pyridine complex of sulfur trioxide provides excellent activation of DMSO in the presence of triethylamine. **The oxidation reaction is usually performed by the addition of the pyridine–sulfur trioxide complex (3 molar equivalents) to a DMSO solution of the alcohol and triethylamine (7–16 molar equivalents) at room temperature.** The addition of triethylamine is presumably necessary to permit efficient proton abstraction from the alkoxysulfonium salt (**130**) giving the ylid (**131**) in the presence of sulfur trioxide and its strongly acidic decomposition products.

A number of steroidal alcohols including testosterone, epitestosterone, cholestanol, and 20α-hydroxypregn-4-en-3-one have been oxidized with notable success, often giving yields of ketones in excess of 80% (*32*). With this reagent mixture, the stereochemistry of the alcohol leads to quite pronounced differences in the efficiency of the oxidation. Thus, as in the case of oxidation by the DMSO–DCC method, the equatorial 11α-hydroxyl group of (**36a**) was readily oxidized in 70% yield to 11-keto-progesterone (**36c**), but the axial 11β-epimer (**36b**) remained inert. This same selectivity permitted direct oxidation of 11β,17α,20α,21-tetra-hydroxypregn-4-en-3-one 21-acetate (**180a**) to hydrocortisone acetate (**180b**) in 90% yield. Similar quite selective oxidation was obtained using DMSO–DCC, but at least four products, including thioethers and 11-keto compounds, were obtained using DMSO–acetic anhydride. One must conclude that the presumed first intermediate (**181**), like the DMSO–DCC adduct (**19**), is considerably more sterically hindered than the acetoxy-dimethylsulfonium ion and cannot readily approach the axial 11β-alcohol. The oxidation of the allylic alcohol (**182a**) to the α,β-unsaturated aldehyde (**182b**) in 70% yield is noteworthy because of the previously mentioned inertness of such compounds using the DMSO–DCC method in the presence of pyridinium trifluoroacetate.

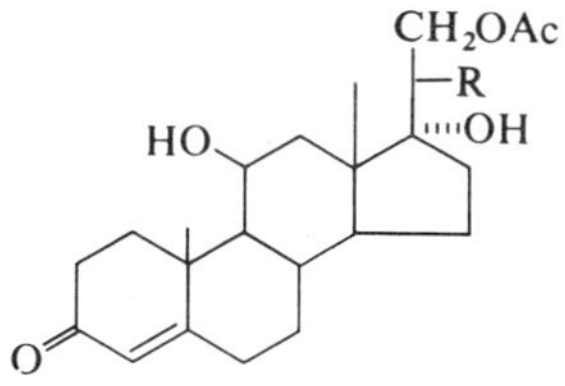

(180)(**a**) R = α-OH, H
(**b**) R = O

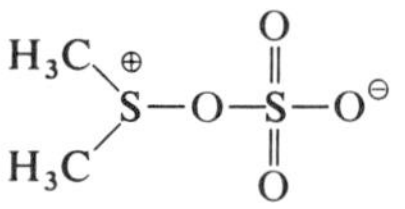

(181)

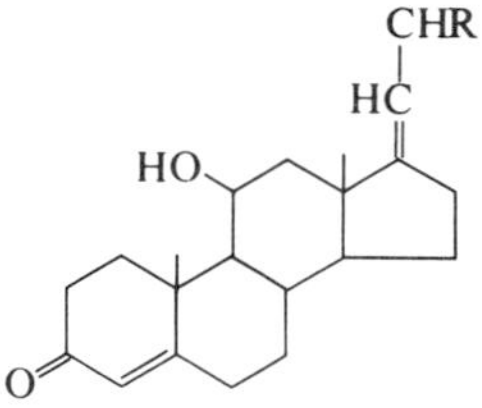

(182)(**a**) R = OH, H
(**b**) R = O

Oxidation of homoallylic alcohols such as 17α-hydroxypregnenolone and 16-dehydropregnenolone are reported to give the unconjugated 5-en-3-ones (*32*). However, these were not isolated as such but rather isomerized in acid to the 4-en-3-ones, which were obtained in yields of 68 and 72%.

The quite strongly basic conditions of the oxidation mixture appears to somewhat limit the use of this method in carbohydrate chemistry. Thus, Cree et al. (*92*) have shown that spontaneous elimination of acetyl groups β to the alcohol being oxidized is a frequent and often efficient process. Treatment of methyl 2,3,4-tri-*O*-acetyl-α-D-mannopyranoside (**183**) gave the α,β-unsaturated aldehyde (**184**) in 75% yield, and a similar result was obtained with 1,2,3,4-tetra-*O*-acetyl-β-D-glucopyranose. In a related way, attempted oxidation of the hemiacetal group of 2,3,4,6-tetra-*O*-acetyl-D-glucopyranose (**185**) was accompanied by elimination of the 3-*O*-acetyl group and formation in 81% yield of the unsaturated lactone (**186**). In the case of 1,3,4,6-tetra-*O*-acetyl-α-D-glucopyranose (**187**), double elimination occurred giving 61% of the dienone (**188**). It may be noted that in all of these cases we appear to be dealing with *cis* eliminations. Since the reaction of 2,3,4,6-tetra-*O*-acetyl-D-mannopyranose (**189**), which is well suited for a *trans* elimination, gives (**186**) at only one-third the rate obtained from (**185**), Cree et al. (*92*) believe that epimerization does not precede elimination. Sugars which do not contain groups that are readily eliminated are efficiently oxidized, 1,2;3,4-di-*O*-isopropylidene-α-D-galactopyranose (**53a**) and 1,2;4,5-di-*O*-isopropylidene-D-fructopyranose (**69**) being converted into the corresponding aldehyde (**53b**) and ketone in yields of 85 and 65%. **The lack of formation of thiomethoxymethyl ethers, even during the preparation of aldehydes, using the sulfur trioxide–DMSO method is striking and makes this an attractive reaction where elimination is not a problem.**

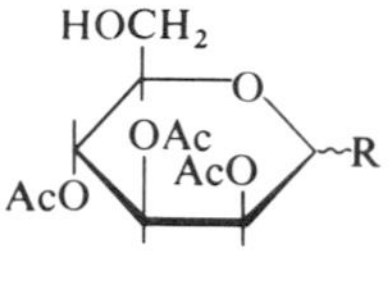

(183) 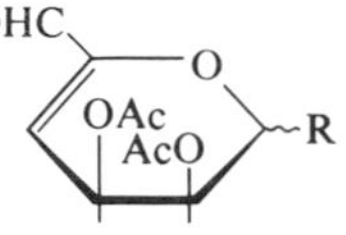 (184) 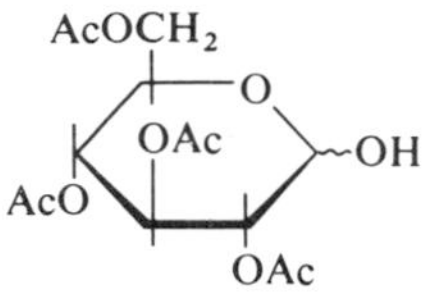 (185) 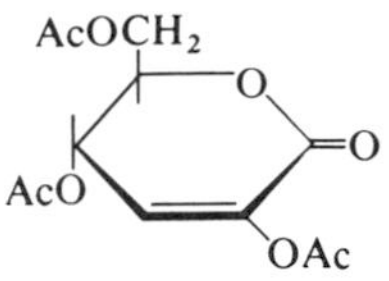 (186)

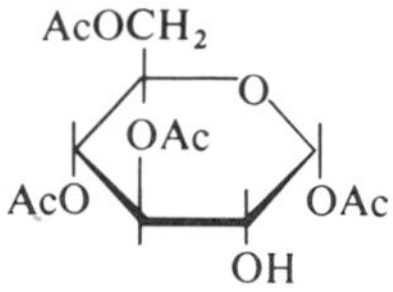

(187) 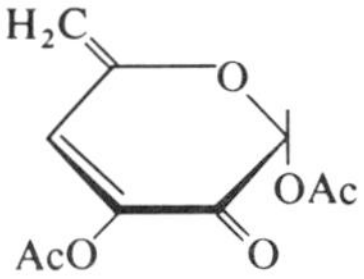 (188) 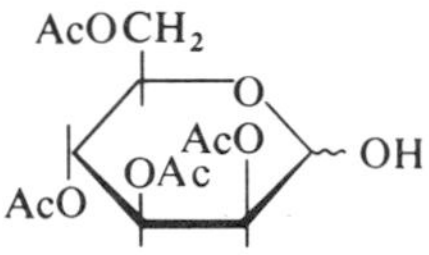 (189)

V. SUMMARY

The development during the past few years of mild reactions in which dimethyl sulfoxide acts as an oxidant has led to the extensive use of these methods in many branches of organic chemistry. The methods which have been considered are those in which DMSO is used in conjunction with an activating agent such as DCC, acetic anhydride, phosphorus pentoxide, and the pyridine complex of sulfur trioxide. All of these reactions have a close mechanistic similarity in that the DMSO is first activated by reaction with the appropriate reagent and subsequent reaction with the alcohol proceeds via intermediates common to all the methods.

To date, perhaps the widest use of these reactions has been in the field of carbohydrate chemistry, where they have met the need for powerful and yet mild oxidative methods. In this chapter the scope and limitations of each method have been considered, and a relatively complete bibliography is provided up to the spring of 1969.

The extensive work done in this laboratory on the reactions of sulfoxides and carbodiimides has been the result of a happy collaboration with many hard-working colleagues. In particular, I would like to acknowledge the devoted efforts of Drs. K. E. Pfitzner, A. H. Fenselau, M. G. Burdon, A. F. Cook, U. Brodbeck, U. Lerch, G. H. Jones, and J. P. H. Verheyden.

REFERENCES

1. D. Martin, A. Weise, and H. J. Niclas, *Angew. Chem. Int. Ed.*, **6,** 318 (1967).
2. C. Agami, *Bull. Soc. Chim. France*, 1021 (1965).
3. J. C. Bloch, *Ann. Chim.*, **10,** 419 (1965).
4. A. J. Parker, *Adv. Org. Chem., Methods and Results*, **5,** 1 (1965).

5. W. W. Epstein and F. W. Sweat, *Chem. Rev.*, **67**, 247 (1967).
6. N. Kornblum, J. W. Powers, G. J. Anderson, W. J. Jones, H. O. Larson, O. Levand, and W. M. Weaver, *J. Amer. Chem. Soc.*, **79**, 6562 (1957).
7. I. M. Hunsberger and J. M. Tien, *Chem. Ind.*, 88 (1959).
8. N. Kornblum, W. J. Jones, and G. J. Anderson, *J. Amer. Chem. Soc.*, **81**, 4113 (1959).
9. H. R. Nace and J. J. Monagle, *J. Org. Chem.*, **24**, 1792 (1959).
10. M. M. Baizer, *J. Org. Chem.*, **25**, 670 (1960).
11. D. N. Jones and M. A. Saeed, *J. Chem. Soc.*, 4657 (1963).
12. R. N. Iacona, A. T. Rowland, and H. R. Nace, *J. Org. Chem.*, **29**, 3495 (1964).
13. T. Cohen and T. Tsugi, *J. Org. Chem.*, **26**, 1681 (1961).
14. T. M. Santosusso and D. Swern, *Tetrahedron Lett.*, 4261 (1968).
15. D. H. R. Barton, B. J. Gardner, and R. H. Wightman, *J. Chem. Soc.*, 1855 (1964).
16. K. H. Scheit and W. Kampe, *Angew. Chem. Int. Ed.*, **4**, 787 (1965).
17. J. Leitich and F. Wessely, *Monatsh. Chem.*, **95**, 129 (1964).
18. K. E. Pfitzner and J. G. Moffatt, Unpublished experiments.
19. G. M. Tener, *J. Amer. Chem. Soc.*, **83**, 159 (1961).
20. K. E. Pfitzner and J. G. Moffatt, *J. Amer. Chem. Soc.*, **87**, 5661 (1965).
21. G. M. Tener, H. G. Khorana, R. Markham, and E. H. Pol, *J. Amer. Chem. Soc.*, **80**, 6223 (1958).
22. K. E. Pfitzner and J. G. Moffatt, *J. Amer. Chem. Soc.*, **85**, 3027 (1963).
23. J. P. Vizsolyi and G. M. Tener, *Chem. Ind.*, 263 (1962).
24. G. P. Moss, C. B. Reese, K. Schofield, R. Shapiro, and Lord Todd, *J. Chem. Soc.*, 1149 (1968).
25. P. Howgate, A. S. Jones, and J. R. Tittensor, *J. Chem. Soc.*, 275 (1968).
26. R. R. Schmidt, V. Schloz, and D. Schwille, *Chem. Ber.*, **101**, 500 (1968).
27. A. S. Jones, R. T. Walker, and A. R. Williamson, *J. Chem. Soc.*, 6033 (1963).
28. A. S. Jones and A. R. Williamson, *Chem. Ind.*, 1624 (1960).
29. R. E. Harmon, C. V. Zenerosa, and S. K. Gupta, *J. Chem. Soc.*, **D**, 327 (1969).
30. J. D. Albright and L. Goldman, *J. Amer. Chem. Soc.*, **89**, 2416 (1967).
31. K. Onodera, S. Hirano, and N. Kashimura, *Carbohyd. Res.*, **6**, 276 (1968).
32. J. R. Parikh and W. von E. Doering, *J. Amer. Chem. Soc.*, **89**, 5505 (1967).
33. U. Lerch, A. H. Fenselau, and J. G. Moffatt, Unpublished experiments.
34. M. G. Burdon and J. G. Moffatt, *J. Amer. Chem. Soc.*, **88**, 5855 (1966).
35. M. G. Burdon and J. G. Moffatt, *J. Amer. Chem. Soc.*, **89**, 4725 (1967).
36. A. F. Cook and J. G. Moffatt, *J. Amer. Chem. Soc.*, **90**, 740 (1968).
37. P. Turnbull, K. Syhora, and J. Fried, *J. Amer. Chem. Soc.*, **88**, 4764 (1966).
38. J. B. Jones and D. C. Wigfield, *Can. J. Chem.*, **44**, 2517 (1966).
39. F. Zetzsche and H. Lindler, *Chem. Ber.*, **71B**, 2095 (1938).
40. H. G. Khorana, *Chem. Rev.*, **53**, 145 (1953).
41. F. Kurzer and K. Douraghi-Zadeh, *Chem. Rev.*, **67**, 107 (1967).
42. J. C. Sheehan and J. J. Hlavka, *J. Org. Chem.*, **21**, 439 (1956).
43. J. C. Sheehan, P. A. Cruikshank, and G. L. Boshart, *J. Org. Chem.*, **26**, 2525 (1961).
44. A. H. Fenselau and J. G. Moffatt, Unpublished results.
45. R. K. Gupta and C. H. Stammer, *J. Org. Chem.*, **33**, 4368 (1968).
46. R. B. Woodward and R. A. Olofson, *J. Amer. Chem. Soc.*, **83**, 1007 (1961).
47. F. Cramer and G. Weimann, *Chem. Ber.*, **94**, 996 (1961).

48. J. F. Arens, *Advan. Org. Chem.*, **2,** 117 (1960).
49. I. Lillien, *J. Org. Chem.*, **29,** 1631 (1964).
50. C. R. Johnson and D. McCant, *J. Amer. Chem. Soc.*, **87,** 5404 (1965).
51. W. von E. Doering and A. K. Hoffman, *J. Amer. Chem. Soc.*, **77,** 521 (1955).
52. S. G. Smith and S. Winstein, *Tetrahedron*, **3,** 317 (1958).
53. G. Cilento, *Chem. Rev.*, **60,** 147 (1960).
54. G. A. Russell and G. J. Mikol, in *Mechanisms of Molecular Migrations* (B. S. Thyragarajan, ed.), Vol. 1, Wiley-Interscience, New York, 1968, p. 157.
55. C. R. Johnson and W. G. Phillips, *J. Amer. Chem. Soc.*, **91,** 682 (1969).
56. A. H. Fenselau and J. G. Moffatt, *J. Amer. Chem. Soc.*, **88,** 1762 (1966).
57. J. G. Moffatt, *J. Org. Chem.*, in Press.
58. K. Torssell, *Acta Chem. Scand.*, **21,** 1 (1967).
59. Unpublished experiments by J. G. Moffatt.
60. F. W. Sweat and W. W. Epstein, *J. Org. Chem.*, **32,** 835 (1967).
61. K. Torssell, *Tetrahedron Lett.*, 4445 (1966).
62. J. N. Cooper and R. E. Powell, *J. Amer. Chem. Soc.*, **85,** 1590 (1963).
63. B. Capon, M. J. Perkins, and C. W. Rees, *Organic Reaction Mechanisms 1967*, Wiley-Interscience, New York, 1968, p. 426.
64a. R. E. Harmon and C. V. Zenarosa, *Abstracts 154th National Meeting of the American Chemical Society*, Chicago, 1967. Abstract D-3.
64b. R. E. Harmon, C. V. Zenarosa, and S. K. Gupta, *Tetrahedron Lett.*, 3781 (1969).
65. K. E. Pfitzner and J. G. Moffatt, *J. Amer. Chem. Soc.*, **87,** 5670 (1965).
66. K. E. Pfitzner, J. P. Marino, and R. A. Olofson, *J. Amer. Chem. Soc.*, **87,** 4658 (1965).
67. C. H. Kuo, D. Taub, and N. L. Wendler, *J. Org. Chem.*, **33,** 3126 (1968).
68. J. Schrieber and A. Eschenmoser, *Helv. Chim. Acta*, **38,** 1529 (1955).
69. F. H. Westheimer and N. Nicolaides, *J. Amer. Chem. Soc.*, **71** (1949).
70. A. Butenandt and J. Schmidt-Thomé, *Chem. Ber.*, **69,** 882 (1936).
71. C. Djerassi, R. R. Engle, and A. Bowers, *J. Org. Chem.*, **21,** 1547 (1956).
72. F. Schneider, J. Hamsker, and R. E. Beyler, *Steroids*, **8,** 553 (1966).
73. J. Carnduff, *Quart. Rev.*, 169 (1966).
74. M. Akhtar, P. F. Hunt, and M. A. Parvez, *Biochem. J.*, **103,** 616 (1967).
75. O. Theander, *Advan. Carbohyd. Chem.*, **17,** 223 (1962).
76. P. J. Beynon, P. M. Collins, P. T. Doganges, and W. G. Overend, *J. Chem. Soc.*, 1131 (1966).
77. H. P. C. Hogenkamp, J. N. Ladd, and H. A. Barker, *J. Biol. Chem.*, **237,** 1950 (1962).
78a. G. H. Jones and J. G. Moffatt, Unpublished experiments.
78b. G. H. Jones and J. G. Moffatt, Abstracts, 158th National Meeting of The American Chemical Society, New York, September 1969, CARB. 16.
79. G. H. Jones and J. G. Moffatt, *J. Amer. Chem. Soc.*, **90,** 5337 (1968).
80. G. H. Jones, J. P. H. Verheyden, and J. G. Moffatt, Abstract N-26 of the XXIst International Congress of Pure and Applied Chemistry, Prague, 1967.
81. P. A. Frey and R. H. Abeles, *J. Biol. Chem.*, **241,** 2732 (1966).
82. H. P. C. Hogenkamp, R. K. Ghambeer, C. Brownson, R. L. Blakley, and E. Vitols, *J. Biol. Chem.*, **243,** 799 (1968).
83. G. B. Howarth, D. C. Lane, W. A. Szarek, and J. K. N. Jones, *Can. J. Chem.*, **47,** 75 (1969).

84. D. Horton, M. Nakadate, and J. M. Tronchet, *Carbohyd. Res.*, **7,** 56 (1968).
85. D. Horton, J. G. Huges, and J. M. Tronchet, *Chem. Commun.*, 481 (1968).
86. H. Saeki and T. Iwashige, *Chem. Pharm. Bull.*, **16,** 1129 (1968).
87. A. Kampf, A. Felsenstein, and E. Dimant, *Carbohyd. Res.*, **5,** 220 (1968).
88. A. F. Cook and J. G. Moffatt, Unpublished results.
89. B. A. Dmitriev, A. A. Kost, and N. K. Kochetkov, *Izv. Akad. Nauk.*, 903 (1969).
90. P. C. Simonart, W. L. Salo, and S. Kirkwood, *Biochem. Biophys. Res. Commun.* **24,** 120 (1966).
91. G. R. Shyrock and H K. Zimmerman, *Carbohyd. Res.*, **3,** 14 (1966).
92. G. M. Cree, D. W. Mackie, and A. S. Perlin, *Can. J. Chem.*, **47,** 511 (1969).
93. J. S. Brimacombe, J. G. H. Bryan, A. Husain, M. Stacey, and M. S. Tolley, *Carbohyd. Res.*, **3,** 318 (1967).
94. Y. Rabinsohn and H. G. Fletcher, *J. Org. Chem.*, **32,** 3452 (1967).
95. H. Fukami, H. S. Koh, T. Sakata, and M. Nakajima, *Tetrahedron Lett.*, 4771 (1968).
96. H. Fukami, H. S. Koh, T. Sakata, and M. Nakajima, *Tetrahedron Lett.*, 1701 (1967).
97. J. R. Dyer, W. E. McGonigal, and K. C. Rice, *J. Amer. Chem. Soc.*, **87,** 654 (1965).
98. B. R. Baker and D. H. Buss, *J. Org. Chem.*, **30,** 2034 (1965).
99. J. J. K. Novák and F. Šorm, *Coll. Czech. Chem. Commun.*, **30,** 3303 (1965).
100. J. Smejkal, J. P. H. Verheyden, and J. G. Moffatt, Unpublished results.
101. E. J. McDonald, *Carbohyd. Res.*, **5,** 106 (1967).
102. K. James, A. R. Tatchell, and P. K. Ray, *J. Chem. Soc.*, **C,** 2681 (1967).
103. G. M. Cree and A. S. Perlin, *Can. J. Biochem.*, **46,** 765 (1968).
104. W. Sowa and G. H. S. Thomas, *Can. J. Chem.*, **44,** 836 (1966).
105. O. Theander, *Acta Chem. Scand.*, **18,** 2209 (1964).
106. B. R. Baker and D. H. Buss, *J. Org. Chem.*, **30,** 2308 (1965).
107. Y. Ali and A. C. Richardson, *Carbohyd. Res.*, **5,** 441 (1967).
108. M. Matsui, M. Saito, M. Okada, and M. Ishadate, *Chem. Pharm. Bull.*, **16,** 1294 (1968).
109. H. Shibata, I. Takashita, N. Kurihara, and M. Nakajima, *Agr. Biol. Chem.*, **32,** 1006 (1968).
110. D. Horton and J. S. Jewell, *Carbohyd. Res.*, **2,** 251 (1966).
111. K. Antonakis, *Bull. Soc. Chim.*, 2972 (1968).
112. M. Delton and G. U. Yuen, *J. Org. Chem.*, **33,** 2473 (1968).
113. H. Kuzuhara and H. G. Fletcher, *J. Org. Chem.*, **32,** 2535 (1967).
114. A. Rosenthal, D. Abson, T. D. Field, H. J. Koch, and R. E. J. Mitchell, *Can. J. Chem.*, **45,** 1525 (1967).
115. A. F. Cook and J. G. Moffatt, *J. Amer. Chem. Soc.*, **89,** 2697 (1967).
116. U. Brodbeck and J. G. Moffatt, *J. Org. Chem.*, **35**, 3552 (1970).
117. K. Bredereck, *Tetrahedron Lett.*, 695 (1967).
118. M. L. Wolfrom and P. Y. Wang, *Chem. Commun.*, 113 (1968).
119. A. N. de Belder, B. Lindberg, and S. Svensson, *Acta Chem. Scand.*, **22,** 949 (1968).
120. R. Bernetti and T. B. Aldrich, *Die Starke*, **20,** 224 (1968).
121. J. D. Albright and L. Goldman, *J. Org. Chem.*, **30,** 1107 (1965).
122. G. Büchi, D. L. Coffen, K. Kocsis, P. E. Sonnet, and F. E. Ziegler, *J. Amer. Chem. Soc.*, **87,** 2074 (1965).
123. R. Arndt and C. Djerassi, *Experientia*, **21,** 566 (1965).
124. H. Achenbach and K. Biemann, *J. Amer. Chem. Soc.*, **87,** 4944 (1965).

125. N. Neuss, L. L. Huckstep, and N. J. Cone, *Tetrahedron Lett.*, 811 (1967).
126. G. Büchi, P. Kulsa, and R. L. Rosati, *J. Amer. Chem. Soc.*, **90,** 2448 (1968).
127. G. Büchi, D. L. Coffen, K. Kocsis, P. E. Sonnet, and F. E. Ziegler, *J. Amer. Chem. Soc.*, **88,** 3099 (1966).
128. W. Nagata, S. Hirai, T. Okumura, and K. Kawata, *J. Amer. Chem. Soc.*, **90,** 1650 (1968).
129. D. F. Waterhouse and B. E. Wallbank, *J. Insect Physiol.*, **13,** 1657 (1967).
130. T. Oishi, M. Nagai and Y. Ban, *Tetrahedron Lett.*, 491 (1968).
131. O. M. Rosen, P. Hoffe, and B. L. Horecker, *J. Biol. Chem.*, **240,** 1517 (1965).
132. S. W. Pelletier and S. Prabhakar, *J. Amer. Chem. Soc.*, **90,** 5318 (1968).
133. H. Zinnes, R. A. Comes, and J. Shavel, *J. Med. Chem.*, 223 (1967).
134. E. J. Corey, K. Lin, and J. Jautelat, *J. Amer. Chem. Soc.*, **90,** 2724 (1968).
135. F. Bohlmann and R. Miethe, *Chem. Ber.*, **100,** 3861 (1967).
136. R. A. Schneider and J. Meinwald, *J. Amer. Chem. Soc.*, **89,** 2023 (1967).
137. J. Karliner and C. Djerassi, *J. Org. Chem.*, **31,** 1945 (1966).
138. K. Weisner, A. Phillips and P. T. Ho, *Tetrahedron Lett.*, 1209 (1968).
139. L. A. Paquette and L. D. Wise, *J. Amer. Chem. Soc.*, **89,** 6659 (1967).
140. P. Dowd and K. Sachdev, *J. Amer. Chem. Soc.*, **89,** 715 (1967).
141. N. M. Weinshenker and F. D. Greene, *J. Amer. Chem. Soc.*, **90,** 506 (1968).
142. A. G. Brook and J. B. Pierce, *J. Org. Chem.*, **30,** 2566 (1965).
143. A. G. Brook, J. M. Duff, P. J. Jones, and N. R. Davis, *J. Amer. Chem. Soc.*, **89,** 431 (1967).
144. A. G. Brook, P. F. Jones, and G. J. D. Peedle, *Can. J. Chem.*, **46,** 2119 (1968).
145. J. W. Huffman, T. Kayima, and C. B. S. Rao, *J. Org. Chem.*, **32,** 700 (1967).
146. R. D. Elliot, C. Temple, and J. A. Montgomery, *J. Org. Chem.*, **33,** 533 (1968).
147. G. Büchi, B. Gubler, R. S. Schneider, and J. Wild, *J. Amer. Chem. Soc.*, **89,** 2776 (1967).
148. E. Klein, W. Rojahn, and D. Henneberg, *Tetrahedron*, **20,** 2025 (1964).
149. E. E. Smissman and W. H. Gastrock, *J. Med. Chem.*, **11,** 860 (1968).
150. E. Klein and W. Rojahn, *Chem. Ber.*, **97,** 2700 (1964).
151. F. D'Angeli, E. Scoffone, F. Filira, and V. Giormani, *Tetrahedron Lett.*, 2745 (1966).
152. F. D'Angeli, V. Giormani, F. Filira, and C. DiBello, *Biochem. Biophys. Res. Commun.*, **28,** 809 (1967).
153. T. Gabriel, W. Y. Chen, and A. L. Nussbaum, *J. Amer. Chem. Soc.*, **90,** 6833 (1968).
154. A. H. Fenselau, E. H. Hamamura, and J. G. Moffatt, *J. Org. Chem.*, **35**, 3546 (1970).
155. T. Tsuyino, *Tetrahedron Lett.*, 4111 (1968).
156. T. Sato, A. Takatsu, Y. Saito, T. Tohyama, and K. Hata, *Bull. Chem. Soc. Japan*, **41,** 221 (1968).
157. A. Guggisberg, M. Hesse, W. von Philipsborn, K. Nagarajan, and H. Schmid, *Helv. Chim. Acta*, **49,** 2321 (1966).
158. J. J. Dugan, M. Hesse, U. Renner, and H. Schmid, *Helv. Chim. Acta*, **50,** 60 (1967).
159. Z. Hori, T. Kurihara, S. Yamamoto, and I. Ninomiya, *Chem. Pharm. Bull.*, **15,** 1641 (1967).
160. S. M. Ifzal and D. A. Wilson, *Tetrahedron Lett.*, 1577 (1967).

161. H. H. Inhoffen, P. Jager, R. Mahlhop, and C. D. Mengler, *Ann. Chem.*, **704,** 188 (1967).
162. G. R. Pettit and S. K. Gupta, *J. Chem. Soc.*, **C,** 1208 (1968).
163. G. Gadsby, M. R. C. Leeming, G. Greenspan, and H. Smith, *J. Chem. Soc.*, **C,** 2647 (1968).
164. H. Kuzuhara and H. G. Fletcher, *J. Org. Chem.*, **32,** 2531 (1967).
165. D. Horton and J. S. Jewell, *Carbohyd. Res.*, **5,** 149 (1967).
166. N. A. Hughes, *Carbohyd. Res.*, **7,** 474 (1968).
167. W. M. zu Reckendorf and J. Feldkamp, *Chem. Ber.*, **101,** 2289 (1968).
168. G. J. F. Chittenden, *Chem. Commun.*, 779 (1968).
169. K. Antonakis, *Bull. Soc. Chim.*, 122 (1969).
170. K. Antonakis and F. Leclercq, *Compt. Rend.*, **265C,** 1004 (1967).
171. W. Sowa, *Can. J. Chem.*, **46,** 1586 (1968).
172. G. L. Tong, W. W. Lee, and L. Goodman, *J. Org. Chem.*, **32,** 1948 (1967).
173. B. Lindberg and K. N. Slessor, *Acta Chem. Scand.*, **21,** 910 (1967).
174. H. P. C. Hogenkamp, *Carbohyd. Res.*, **3,** 239 (1966).
175. M. Kawana, H. Ohrui, and S. Emoto, *Bull. Chem. Soc. Japan*, **41,** 2199 (1968).
176. J. S. Brimacombe and A. Hussain, *Carbohyd. Res.*, **6,** 491 (1968).
177. M. Cerny, L. Kalvoda, and J. Pacak, *Coll. Czech. Chem. Commun.*, **33,** 1143 (1968).
178. A. Rosenthal and P. Calsoulacos, *Can. J. Chem.*, **46,** 2868 (1968).
179. O. Gabriel, *Carbohyd. Res.*, **6,** 319 (1968).
180. E. Zissis, *J. Org. Chem.*, **32,** 660 (1967).
181. D. E. Kiely and H. G. Fletcher, *J. Org. Chem.*, **33,** 3723 (1968).
182. K. Antonakis, F. Leclercq, and M. J. Arvor, *Compt. Rend.*, **264,** 524 (1967).
183. E. Zissis, *J. Org. Chem.*, **33,** 2844 (1968).
184. A. J. Fatiadi, *Chem. Commun.*, 441 (1967).
185. G. E. McCasland, M. O. Nauman, and L. J. Durham, *J. Org. Chem.*, **33,** 4220 (1968).
186. J. L. Godman and D. Horton, *Carbohyd. Res.*, **6,** 229 (1968).
187. K. T. Joseph and C. S. Krishna Rao, *Tetrahedron*, **23,** 3215 (1967).
188. E. Eggart, C. Pascual, and H. Wehrli, *Helv. Chim. Acta*, **50,** 985 (1967).
189. J. W. Clark-Lewis and D. C. Skingle, *Aust. J. Chem.*, **20,** 2169 (1967).
190. B. E. Cross and P. L. Myers, *J. Chem. Soc.*, **C,** 471 (1968).
191. B. Houghton and J. E. Saxton, *Tetrahedron Lett.*, 5475 (1968).
192. V. K. Bhatia, H. C. Krishnamurty, R. Madhav, and T. R. Seshadri, *Tetrahedron Lett.*, 3859 (1968).
193. M. S. Newman and C. C. Davis, *J. Org. Chem.*, **32,** 66 (1967).
194. M. VanDyke and N. K. Pritchard, *J. Org. Chem.*, **32,** 3204 (1967).
195. J. J. Bloomfield, J. R. S. Ireland, and A. P. Marchand, *Tetrahedron Lett.*, 5647 (1968).
196. H. W. Moore and R. J. Wikholm, *Tetrahedron Lett.*, 5049 (1968).
197. K. Weinges and F. Nader, *Ann. Chem.*, **706,** 112 (1967).
198. T. Kabota, S. Matsutani, and M. Shiro, *Chem. Commun.*, 1541 (1968).
199. K. D. Barrow, D. H. R. Barton, E. B. Chain, U. F. W. Ohnsorge, and R. Thomas *Chem. Commun.*, 1198 (1968).
200. R. Nagarajan, L. L. Huckstep, D. H. Lively, D. C. deLong, M. M. March, and N. Neuss, *J. Amer. Chem. Soc.*, **90,** 2980 (1968).
201. Y. Hayashi and R. Oda, *J. Org. Chem.*, **32,** 457 (1967).

202. G. R. Pettit and T. H. Brown, *Can. J. Chem.*, **45,** 1306 (1967).
203. P. Claus, *Monatsh. Chem.*, **99,** 1034 (1968).
204. A. Hochrainer and F. Wessely, *Monatsh. Chem.*, **97,** 1 (1966).
205. H. Nozaki, Z. Morita, and K. Kondo, *Tetrahedron Lett.*, 2913 (1966).
206. K. Onodera, S. Hirano, and N. Kashimura, *J. Amer. Chem. Soc.*, **87,** 4651 (1965).
207. W. M. zu Reckendorf, *Chem. Ber*, **101,** 3802 (1968).
208. C. L. Stevens, R. P. Glinski, and K. G. Taylor, *J. Org. Chem.*, **33,** 1586 (1968).
209. W. M. zu Reckendorf, *Chem. Ber.*, **101,** 3652 (1968).
210. A. Rosenthal and D. A. Baker, *Tetrahedron Lett.*, 397 (1969).

2

Photosensitized Oxygenations

*W. R. ADAMS**
UNION CARBIDE CORPORATION
CHEMICALS AND PLASTICS
SOUTH CHARLESTON, WEST VIRGINIA

I. Introduction 65
II. Methods of Generating Singlet Oxygen. 69
III. General Techniques 71
IV. Photooxygenation of Conjugated Dienes 77
V. Photooxygenation of Heterocyclic Pentadiene Derivatives . . 81
A. Furans. 81
B. Pyrroles 84
C. Oxazoles 86
D. Imidazoles 88
E. Thiophenes 91
VI. Photooxygenation of Acenes 92
VII. Photooxygenation of Olefins 94
A. Monoenes and Dienes 94
B. Polyolefins 100
References 109

I. INTRODUCTION

The first dye-sensitized photooxygenation reaction was discovered in 1928 by Windaus and Brunken (*1*) with the isolation of ergosterol endoperoxide. Since this initial discovery, the reaction has been studied extensively by many investigators. In the presence of suitable organic acceptors, sensitizers (dyes or aromatic hydrocarbons), light, and oxygen, a reaction occurs to afford a mild and efficient method for introducing molecular oxygen in a highly specific fashion into organic compounds. As a class, the reactions proceed smoothly and are preparatively useful because of their high yields.

There are three principal types of reactions. The first type involves the photooxygenation of polycyclic aromatic hydrocarbons, conjugated cyclic dienes, and many heterocycles to give cyclic peroxides. The reaction

* Present address: Sun Chemical Corporation, Corporate Research Division, Carlstadt, New Jersey.

is analogous to a photo-induced Diels–Alder reaction in which the peroxide is formed by a 1,4-addition of oxygen as shown in Eq. (1).

$$\text{cyclopentadiene} + O_2 \xrightarrow[\text{sens.}]{h\nu} \text{2,3-dioxabicyclo[2.2.1]hept-5-ene} \qquad (1)$$

The second type is the photooxygenation of olefins that contain at least one allylic hydrogen to yield allylic hydroperoxides in which the double bond has shifted to a position adjacent to the original double bond [Eq. (2)].

$$-\overset{1}{\underset{|}{C}}=\overset{2}{C}-\overset{3}{\underset{\underset{H}{|}}{C}} + O_2 \xrightarrow[\text{sens.}]{h\nu} \overset{1}{\underset{\underset{O-OH}{|}}{C}}-\overset{2}{C}=\overset{3}{C} \qquad (2)$$

This reaction bears a formal resemblance to the "ene" reaction depicted in Eq. (3).

$$\text{propene (allylic H)} + \text{maleic anhydride} \longrightarrow \text{allylsuccinic anhydride} \qquad (3)$$

More recently a third general class of reaction was discovered that involves the combination of singlet oxygen with "reactive double bonds" to give 1,2-dioxetanes as intermediates. These intermediates appear to be responsible for carbonyl fragmentation products in the reactions of singlet oxygen with olefins [Eq. (4)] (*2*).

$$R_2C=CR_2 + {}^1O_2 \longrightarrow \left[\begin{array}{c} O-O \\ R_2C-CR_2 \end{array}\right] \longrightarrow 2R_2C=O \qquad (4)$$

It is essential from the onset that the sensitized photooxygenations be distinguished from the more familiar radical oxidations. For example, the radical chain autoxidation of an olefin involves the initial thermal or photolytic homolysis of a free-radical initiator that abstracts the allylic hydrogen atom from the olefin giving the intermediate allylic radical. This step is subsequently followed by molecular oxygen addition to the allylic radical. Formation of the hydroperoxide and a second allylic radical completes the cycle as shown in Scheme 1.

Scheme 1

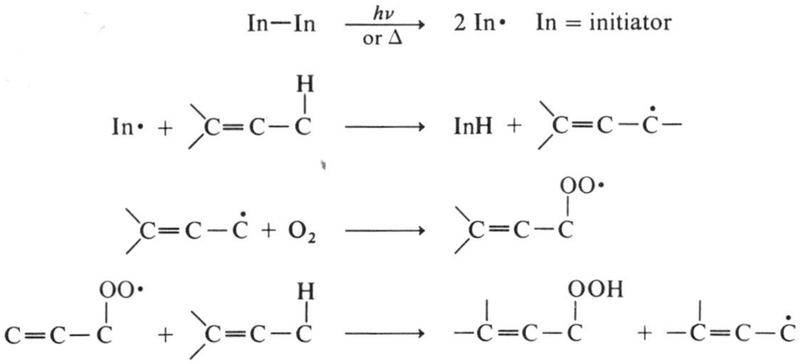

$$\text{In—In} \xrightarrow[\text{or } \Delta]{h\nu} 2\,\text{In}\cdot \qquad \text{In} = \text{initiator}$$

$$\text{In}\cdot + \;>\!C{=}C{-}\overset{\overset{\displaystyle H}{|}}{C} \longrightarrow \text{InH} + \;>\!C{=}C{-}\dot{C}{-}$$

$$>\!C{=}C{-}\dot{C} + O_2 \longrightarrow \;>\!C{=}C{-}\overset{\overset{\displaystyle OO\cdot}{|}}{C}$$

$$C{=}C{-}\overset{\overset{\displaystyle OO\cdot}{|}}{C} + \;>\!C{=}C{-}\overset{\overset{\displaystyle H}{|}}{C} \longrightarrow {-}\overset{|}{C}{=}C{-}\overset{\overset{\displaystyle OOH}{|}}{C} + {-}\overset{|}{C}{=}C{-}\dot{C}$$

The benzophenone-sensitized photooxygenation of isopropyl alcohol (*3*) is typical of a second type of radical oxidation in which the electronically excited benzophenone (S) initiates the oxidation by abstracting a hydrogen atom from isopropyl alcohol. The initiation is then followed by molecular oxygen addition to the 2-hydroxypropyl radical. In the termination step the 2-hydroperoxy-2-propyl radical abstracts a hydrogen atom from the semireduced sensitizer to regenerate a ground state molecule of benzophenone (S) (Scheme 2). Although isopropyl alcohol is a very reactive

Scheme 2

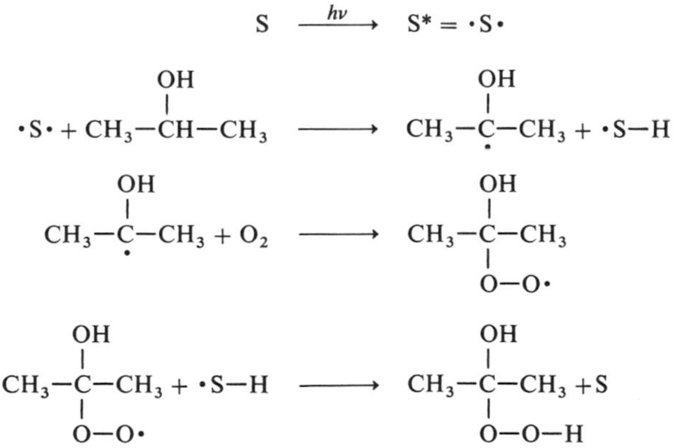

$$S \xrightarrow{h\nu} S^* = \cdot S \cdot$$

$$\cdot S \cdot + CH_3{-}\overset{\overset{\displaystyle OH}{|}}{CH}{-}CH_3 \longrightarrow CH_3{-}\underset{\bullet}{\overset{\overset{\displaystyle OH}{|}}{C}}{-}CH_3 + \cdot S{-}H$$

$$CH_3{-}\underset{\bullet}{\overset{\overset{\displaystyle OH}{|}}{C}}{-}CH_3 + O_2 \longrightarrow CH_3{-}\underset{\underset{\displaystyle O{-}O\cdot}{|}}{\overset{\overset{\displaystyle OH}{|}}{C}}{-}CH_3$$

$$CH_3{-}\underset{\underset{\displaystyle O{-}O\cdot}{|}}{\overset{\overset{\displaystyle OH}{|}}{C}}{-}CH_3 + \cdot S{-}H \longrightarrow CH_3{-}\underset{\underset{\displaystyle O{-}O{-}H}{|}}{\overset{\overset{\displaystyle OH}{|}}{C}}{-}CH_3 + S$$

acceptor for radical oxidations, it is inactive in dye-sensitized oxidations.

Each of these oxidation processes involves a reaction of radical intermediates with a ground state, or triplet electronic state, oxygen molecule. Triplet oxygen is paramagnetic, having two unpaired electrons with parallel spins, making it especially amenable to free-radical reactions.

The dye-sensitized photooxygenation, in contrast, is a nonchain reaction and involves only electronically excited states as intermediates. The reaction is represented schematically in Scheme 3.

Scheme 3

$$\text{Sens.} \xrightarrow{h\nu} {}^1\text{Sens.}$$
$${}^1\text{Sens.} \longrightarrow {}^3\text{Sens.}$$
$${}^3\text{Sens.} + {}^3O_2 \longrightarrow \text{Sens.} + {}^1O_2$$
$${}^1O_2 + \text{RH} \longrightarrow \text{ROOH} \qquad \text{RH} = \text{olefin}$$

The oxidation involves the absorption of light by the sensitizer which undergoes an electronic transition to the excited singlet state (^{1}Sens.). Following an efficient inter-system crossing from the singlet state to the excited triplet state of the sensitizer, energy is transferred to ground state molecular oxygen to yield an excited singlet state of molecular oxygen. Singlet oxygen then reacts with the appropriate organic substrate to give the oxidized product (peroxide or hydroperoxide).

The mechanism of the photooxygenation reaction was originally proposed by Kautsky (*4*). More recently, Sharp (*5*) and Foote and Wexler (*6,7*) have suggested that the reactive intermediate is an excited singlet state of molecular oxygen. Molecular oxygen has two metastable singlet states with spectroscopic symmetry notations $^1\Sigma_g^+$ and $^1\Delta_g$ (Table I). The

TABLE I

Electronic States and Configurations of Oxygen Molecule

State	Occupancy of highest orbitals		Energy, kcal
$^1\Sigma_g^+$	↑	↓	37
$^1\Delta_g$	↑ ↓		22
$^3\Sigma_g^-$	↑	↑	Ground state

lifetime of the electronic states are quite different: the $^1\Delta_g$ oxygen lifetime has been estimated to be approximately 1 hr (*8,9*), and estimates for $^1\Sigma_g^+$ oxygen are less than 10^{-1} sec (*10*). Kearns and coworkers (*11,12*) have proposed that both $^1\Sigma_g^+$ and $^1\Delta_g$ oxygen molecules are involved as intermediates in the photooxygenation reaction.

In this chapter the author does not intend to include a comprehensive review of the literature. Several excellent reviews concerned with the scope and the mechanism of the photooxygenation reaction are available (*13–20*). Rather, the intent is to stress the usage of the reaction from a preparative point of view, with particular regard to the substrates used and products formed.

II. METHODS OF GENERATING SINGLET OXYGEN

In addition to the classical photosensitization method for generating singlet oxygen, several alternative procedures have been developed recently. For example, singlet oxygen can be produced by subjecting gaseous oxygen to an electrodeless discharge (*21,22*). The apparatus consists of a radiofrequency unit with output leads attached to two aluminum foil bands fitted around quartz tubing. Anthracene, 9,10-dimethylanthracene, and 9,10-diphenylanthracene have been thus converted to the corresponding endoperoxides, identical to samples prepared by the photooxygenation routes.

Murray and Kaplan (*23,24*) have shown that the adduct of triphenyl phosphite and ozone prepared at $-70°$ can be used as a source of singlet oxygen (Scheme 4). This method affords a means whereby the excited oxygen can be released at a carefully controlled rate.

Scheme 4

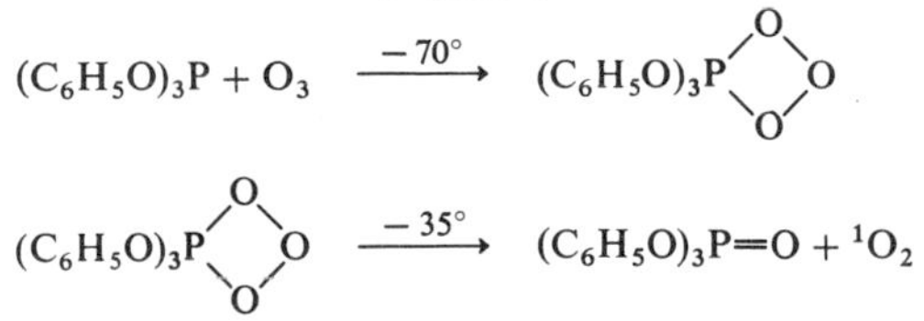

$$(C_6H_5O)_3P + O_3 \xrightarrow{-70°} (C_6H_5O)_3P\langle O_3 \rangle$$

$$(C_6H_5O)_3P\langle O_3 \rangle \xrightarrow{-35°} (C_6H_5O)_3P{=}O + {}^1O_2$$

9,10-Diphenylanthracene-9,10-peroxide (**1**) may be used to bring about singlet oxygen reactions when it is allowed to decompose in the presence of a variety of known singlet oxygen acceptors (*25*).

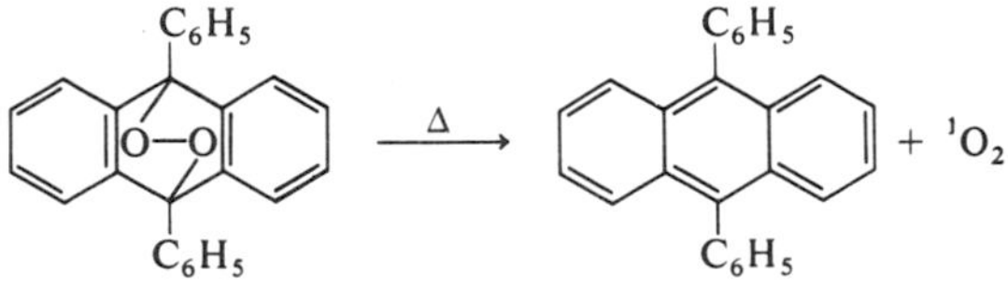

C_6H_5 C_6H_5

O—O $\xrightarrow{\Delta}$ $+ {}^1O_2$

C_6H_5 C_6H_5

(**1**)

The regeneration of cyclohexanone from either the thermolysis or the photolysis of the ketone peroxide (**2**) yields molecular oxygen (probably singlet) and two molecules of ketone (*26*). However, this procedure is more of theoretical than practical significance, since the conversion to singlet oxygen is very inefficient.

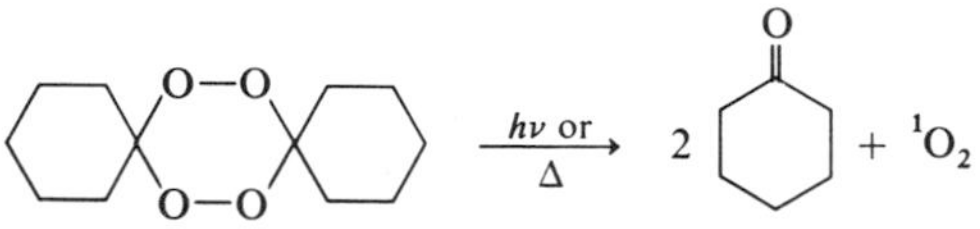

$$\text{(2)} \xrightarrow[\Delta]{h\nu \text{ or}} 2 \text{ cyclohexanone} + {}^1O_2$$

Chemiluminescence was observed when benzene solutions of benzoyl peroxide and cumene hydroperoxide were heated to 80° (*27*). However, the peroxide decomposition was either too slow or the yield of oxygen was too low to be of real synthetic value.

Singlet oxygen has also been generated from the ceric ion oxidation of *sec*-butyl hydroperoxide (*28*). The excited oxygen arises from the combination of *sec*-butylperoxy radicals forming a tetroxide which subsequently decomposes as shown in Scheme 5.

Scheme 5

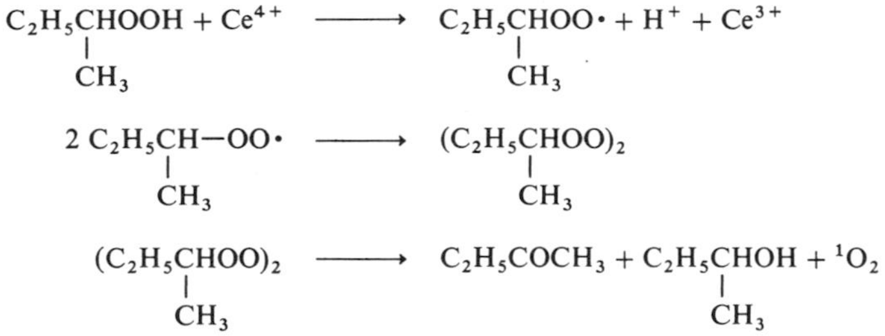

$$C_2H_5CH(CH_3)OOH + Ce^{4+} \longrightarrow C_2H_5CH(CH_3)OO\cdot + H^+ + Ce^{3+}$$

$$2\,C_2H_5CH(CH_3){-}OO\cdot \longrightarrow (C_2H_5CH(CH_3)OO)_2$$

$$(C_2H_5CH(CH_3)OO)_2 \longrightarrow C_2H_5COCH_3 + C_2H_5CH(CH_3)OH + {}^1O_2$$

Chemically, singlet oxygen is formed *in situ* by the heterolytic decomposition of hydrogen peroxide. Foote and Wexler (*6*) have carefully studied the reaction between sodium hypochlorite and hydrogen peroxide and have found that this method of producing singlet oxygen parallels the dye-sensitized oxygenations in every respect. The oxidations proceed smoothly by this apparently practical synthetic method.

Excited singlet oxygen has also been shown to be formed in solution from the reaction of alkaline hydrogen peroxide with hypochlorous acid, chlorine, or bromine (*29–31*). For example, addition of hydrogen peroxide and bromine to a solution of polycyclic aromatic hydrocarbons in alkaline potassium hydroxide yields intermediates comparable to those obtained from the photooxygenation reaction but which undergo subsequent solvolysis in the reaction medium (*32*).

Another procedure which has been shown spectroscopically to evolve singlet oxygen is the so-called Trautz reaction (*33,34*). This involves the oxidation of formaldehyde by alkaline hydrogen peroxide in the presence of polyhydric phenols (usually pyrogallol). McKeown and Waters (*32*) found that resorcinol could be used in place of pyrogallol to give an increased yield of the endoperoxide from 9-methyl-10-phenylanthracene.

Several other reactions also appear to generate singlet oxygen; for example, the decomposition of diisoperoxyphthalic acid in alkaline solution (*32*), the reaction between *p*-peroxytoluic acid and alkaline hydrogen peroxide (*32*), and the hydrolysis of nitriles employing hydrogen peroxide (*32*).

For preparative work, the oxygenation with hydrogen peroxide and sodium hypochlorite and the dye-sensized photooxygenation show the greatest promise. The chemical procedure offers a convenient and practical synthetic method for oxidizable substrates that give high quantum yields in the photosensitized oxygenation (e.g., furans, imidazoles, tetra- and trisubstituted olefins). However, where the parallel photosensitized oxidation proceeds in low quantum yield, a large excess of reagents is required. This disadvantage is easily overcome in the photochemical procedure by simply increasing the irradiation time.

III. GENERAL TECHNIQUES

There have been elaborate descriptions of instrumentation developed for both preparative and kinetic work published in the literature (*13,35*). Photochemical reactors for preparative work are readily improvised and consist generally of a water-jacketed Pyrex immersion well fitted into a jacketed cylindrical vessel with a 55/50 standard taper ground-glass joint (Fig. 1). The vessel is equipped with a side arm and a 24/40 standard taper ground-glass joint for a reflux or Dry Ice condenser, a thermowell, and a ball socket joint fitted at the base of the vessel equipped with a glass frit and two stopcocks. The volume of the vessel when enclosing the immersion well is approximately 2.5 liters. Light sources generally employed for photooxygenations are tungsten filament lamps and mercury and xenon medium- to high-pressure radiant lamps. The reaction consists of a substrate, a dye, and in most cases, a solvent.

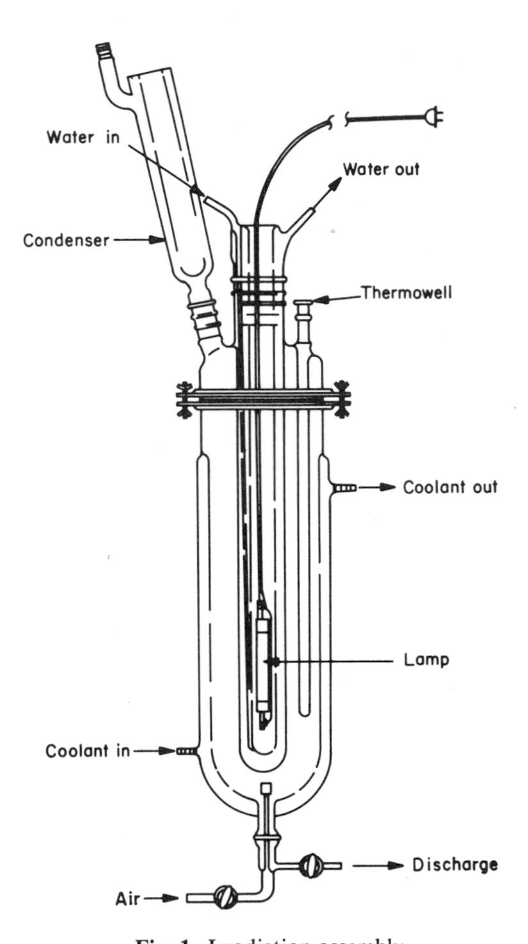

Fig. 1. Irradiation assembly.

Aromatic ketones (e.g., benzophenone and xanthone) and a variety of dyes have been shown to be effective photosensitizers. However, the most useful sensitizers for synthetic operations are the xanthene dyes (e.g., fluorescein, eosin, erythrosin B, and rose bengal), the azine dyes (e.g., methylene blue), and certain porphyrins. The more efficient sensitizers have strong absorptions in the visible region of the spectrum (Table II) and give long-lived triplet states in high quantum yield. The energy of these triplet states lies in the range of 30–50 kcal/mole for maximum sensitization.

TABLE II

Ultraviolet Absorption and Triplet Energy of Various Dye Sensitizers

Dye	λ_{max}, mμ[a]	E_t, kcal/mole
Fluorescein	480, 453	50
Rose bengal	557, 522	39.4 (45.4)
Methylene blue	653	33
		>40
Acridine orange	488	
Acridine yellow	455	
Eosin	545, 500 (sh)	42.4
Erythrosin B		42.0
Riboflavin	447, 350	
Rhodamine B	542	
Rhodamine 6G	529	
Acridine orange	492	

[a] The ultraviolet spectra were measured in 95% ethanol.

A series of dyes were tested for their ability to sensitize the oxidation of levopimaric acid (**3**) (Table III) (*36*). Certain dyes, such as thymol blue, crystal violet, rosaniline, and phenolphthalein-type dyes, displayed no activity whatever.

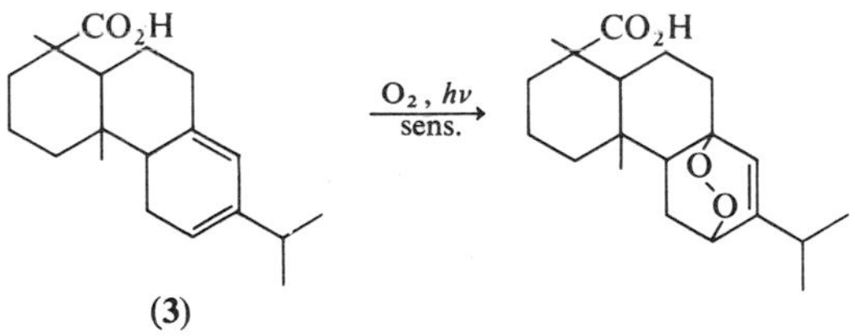

(**3**)

Photosensitized oxygenation of Δ^4-cholesten-3β-ol (**4**) produces a 5α-hydroperoxide that subsequently decomposes to 4α,5-epoxy-5α-cholesten-3-one (**5**) and Δ^4-cholesten-3-one (**6**) (*37,38*). Kearns and co-workers (*11,12*) have demonstrated that variation of sensitizer markedly alters the ratio of enone to epoxy ketone. Using low-energy sensitizers such as methylene blue, these workers observed the resulting ketone/epoxide ratio to be 1 : 3. This ratio was completely reversed by employing

TABLE III

Effect of Sensitizer on the Photooxidation of Levopimaric Acid (**3**)

Sensitizer	Time required to complete reaction
Rose bengal	3.5
Methylene blue	4.5
Erythrosin	4.8
Eosin	7.7
Chlorophyl	72
2,5-Dimethyl-*p*-quinone	100

such higher energy sensitizers as fluorescein or eosin Y. Under these conditions a ketone/epoxide ratio of 3 : 1 was obtained.

HO, H (**4**) $\xrightarrow[\text{sens.}]{h\nu,\ O_2}$ [HO, OOH] ; $-H_2O$ → (**5**) ; $-H_2O_2$ → (**6**)

Due to the high extinction coefficients of these sensitizers, only a small amount of dye is needed to induce a photooxygenation. **Typically, 50–100 mg/110 g of substrate is sufficient for preparative use.**

The choice of solvent usually depends on the mutual solubility of substrate and dye in the medium. Solvents most commonly employed are alcohols (methanol, ethanol, and isopropyl alcohol), tetrahydrofuran, water, pyridine, chloroform, methylene chloride, and carbon disulfide. A remarkable strong influence of the solvent on the rate of oxygen uptake and subsequent peroxide formation was first observed by Dufraisse and Badoche (*39,40*). The rate of photooxidation of rubrene (**7**) was found to be enhanced when carbon disulfide or chloroform was used as the reaction solvent (Table IV).

C_6H_5 C_6H_5 / C_6H_5 C_6H_5 (7) $+ O_2 \xrightarrow{h\nu}$ C_6H_5 C_6H_5 (O–O) C_6H_5 C_6H_5

TABLE IV

Effect of Solvent on the Rate of Photooxidation of Rubrene (7)

Solvent	Relative rate of peroxide formation
Carbon disulfide	9
Chloroform	3
Methyl iodide	1
Benzene	1
Acetone	1
Ethyl ether	0.5
Pyridine	0.25
Nitrobenzene	0.1
Carbon disulfide (75%) and ethyl ether (25%)	2

This effect has been studied extensively by Bowen (*41–43*) with anthracene and some of its derivatives. When a carbon disulfide solution of anthracene is irradiated in the presence of oxygen, the transannular peroxide is the only product formed. However, if benzene is used as the solvent, the peroxide formation is accompanied by dimerization of the anthracene. The quantum yields of peroxide formation were 0.70 and 0.13 in carbon disulfide and benzene, respectively. Since dimerization can only occur from the reaction of a singlet-excited anthracene with a ground state anthracene molecule, the role of carbon disulfide is probably that of a promoter of singlet–triplet intersystem crossing. This proposal is supported by the observation that in carbon disulfide no fluorescence of anthracene is detectable, whereas fluorescence is observed in benzene solutions (*44*).

Useful combinations of solvents have been devised to take advantage of the efficacy of carbon disulfide. For example, Forbes and Griffiths (*45*) have reported a new solvent system comprising of a mixture of carbon disulfide, methanol, and ether in the ratio of 14 : 1 : 1.5 (v/v/v), respectively. This new solvent system is of general application to most of the photooxidizable substrates. The rates of oxidation were observed to be greatly enhanced, with generally higher yields than those obtained in other solvents.

A factor of prime importance in photooxygenation reactions is the structure of the substrate. Kopecky and Reich (*46*) have measured the relative rates of photooxidation, sensitized with methylene blue, of a number of olefins. The most highly substituted olefin studied, 2,3-dimethyl-2-butene, is oxidized 5500 times faster than cyclohexene. Thus, a tetrasubstituted olefin will be attacked preferentially to a trisubstituted olefin in the same molecule, while a mono- or disubstituted olefin will remain essentially inert.

The relative reactivities of a variety of acceptors have been determined by Foote and coworkers (*16*). The relative reactivities cover a span of five powers of ten, with 2,5-dimethylfuran being the most reactive of the substrates investigated (Table V).

TABLE V

Relative Reactivities of Acceptors

Acceptor	Photooxidation k (relative)
2,5-Dimethylfuran	2.4
Cyclopentadiene	1.2
2,3-Dimethyl-2-butene	(1.00)
1,3-Cyclohexadiene	0.08
1-Methylcyclopentene	0.05
trans-3-Methyl-2-pentene	0.04
cis-3-Methyl-2-pentene	0.03
2-Methyl-2-butene	0.024
2-Methyl-2-pentene	0.019
1-Methylcyclohexene	0.0041
cis-4-Methyl-2-pentene	0.00026
Cyclohexene	0.000048

An excellent example that displays the influence of electron density on reactivity is provided by the photooxygenation of dimethylcyclohexa-3,5-diene-1,2-dicarboxylate (**8**). Consistent with the electrophylic character of singlet oxygen, the formation of endoperoxide occurred only with (**8**); no reaction whatsoever took place with (**9**) (*47*).

CO_2CH_3 / CO_2CH_3 (**8**) $\xrightarrow[\text{sens.}]{h\nu,\ O_2}$ O-O endoperoxide with CO_2CH_3 / CO_2CH_3

CO_2CH_3 / CO_2CH_3 (**9**) $\xrightarrow[\text{sens.}]{h\nu,\ O_2}$ No reaction

Side reactions, such as free-radical oxygenation, are more prone to occur with substrates possessing double bonds that are relatively unreactive to singlet oxygen. In such cases, it is advisable to add a free-radical quencher (hydroquinone) to the solution prior to irradiation.

IV. PHOTOOXYGENATION OF CONJUGATED DIENES

The dye-sensitized photooxygenation of conjugated cyclic dienes yields cyclic peroxides formed by a 1,4-addition of oxygen. For example, photooxidation of 1,3-cyclohexadiene, with methylene blue as sensitizer, yields 5,6-dioxabicyclo[2.2.2]octene-2 (**10**) (*48*). The structure of the endo-peroxide was elucidated by hydrogenation in the presence of platinum catalyst giving *cis*-cyclohexane-1,4-diol (**11**). Reduction of (**10**) with

$\xrightarrow[\text{sens.}]{h\nu,\ O_2}$ (**10**) $\xrightarrow{\phi_3P}$ (**13**)

(**10**) $\xrightarrow{H_2,\ Pt}$ (**11**) OH / OH

(**10**) $\xrightarrow{\text{thiourea}}$ (**12**) OH / OH

(**12**) $\xrightarrow[\text{Pd/BaSO}_4]{H_2}$ (**11**)

thiourea yielded 2-cyclohexen-*cis*-1,4-diol (**12**), which was further reduced to (**11**) by hydrogenation in the presence of palladium and barium sulfate. 1,4-Oxa-2-cyclohexene (**13**) was obtained by treating the endoperoxide (**10**) with triphenylphosphine.

In a similar manner, α-terpinene (**14**) yields ascaridole (**15**) in excellent yield (*49*). The chlorophyll-catalyzed synthesis of ascaridole is used commercially in Europe (*50*).

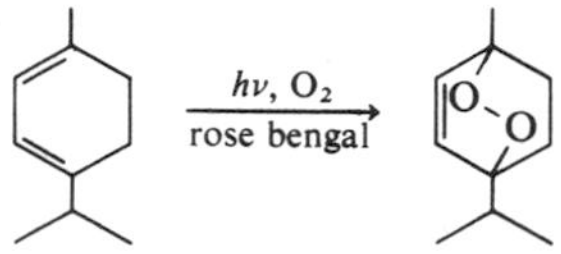

$h\nu$, O_2, rose bengal

(**14**) (**15**)

α-Phellandrene (**16**) yields two isomeric transannular peroxides (52% yield) resulting from the attack of oxygen on either side of the ring (*49,51*).

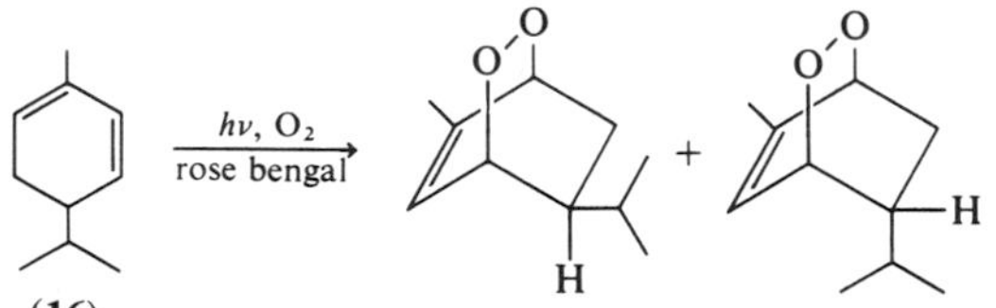

$h\nu$, O_2, rose bengal

(**16**)

Schenck and Wirtz (*52*) have utilized the photooxidation reaction in the synthesis of cantharidin (**17**), wherein the initial oxidation step yields the transannular peroxide (**18**) in 50% yield.

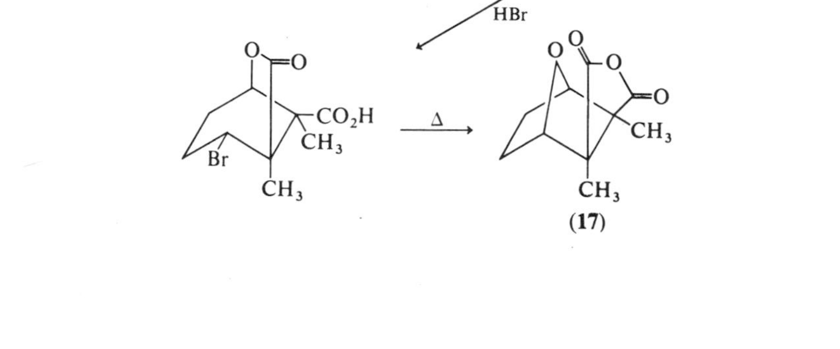

$h\nu$, O_2, rose bengal; H_2, PtO_2; HBr; Δ

(**18**) (**17**)

Schenck and Dunlap (*48*) carried out the dye-sensitized photooxidation of cyclopentadiene at temperatures between −20 and −130°, producing the transannular peroxide (**19**) in excellent yield. The cyclic peroxide was found to be very unstable and was further characterized by reduction to *cis*-1,3-cyclopentadiol (**20**).

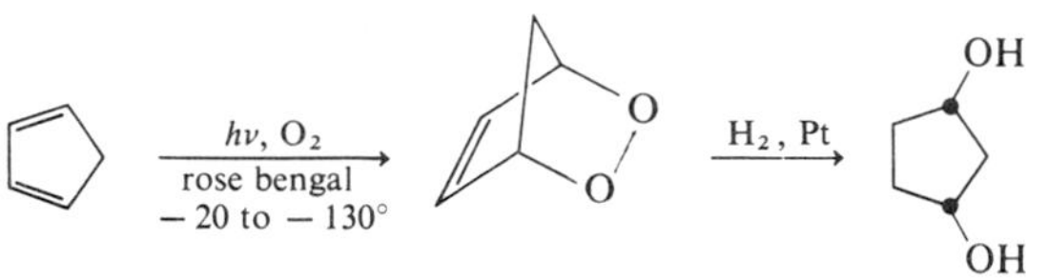

$h\nu$, O_2, rose bengal, −20 to −130° → (**19**) 86% → H_2, Pt → (**20**)

At ambient temperatures the photooxidation allegedly afforded 4-hydroxy-2-cyclopentenone (**23**) (*48,53,54*). Recently, the photooxidation of cyclopentadiene was reinvestigated (*55,56*) utilizing temperatures varying from 10 to 25° in an effort to retard the dimerization of the diene substrate. It was found, contrary to Schenck's report, that 4,5-epoxy-*cis*-2-pentenal (**21**) and *cis*-1,2:3,4-diepoxycyclopentane (**22**) were the major products. No trace of the hydroxyenone (**23**) was observed.

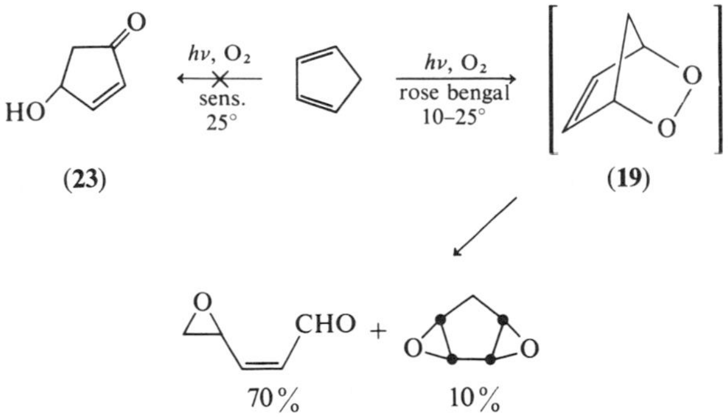

(**23**) ←✕— $h\nu$, O_2, sens., 25° — cyclopentadiene — $h\nu$, O_2, rose bengal, 10–25° → [(**19**)] → (**21**) 70% + (**22**) 10%

However, when the photooxidation was carried out in alkaline solutions, the hydroxyenone (**23**) was obtained in 30% yield (*56*).

Both products presumably arise from a thermal isomerization of the cyclic peroxide. The formation of (**21**) from (**19**) represents a novel four-center peroxide rearrangement.

Substitution of cyclopentadiene at C-1 and C-4 with phenyl groups stabilizes the resulting 1,4-diphenyl-1,4-endoperoxy-2-cyclopentene (**24**) (*57*).

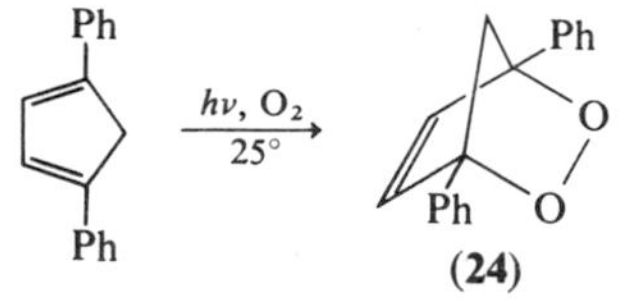

$\xrightarrow[25°]{h\nu,\ O_2}$ (**24**) 60%

1,4-Diphenylcyclopentadiene, which is highly fluorescent, is capable of undergoing a direct photooxygenation. However, the yield of peroxide is markedly enhanced when the photooxidation takes place in the presence of a sensitizer. The sensitizers, in addition to their energy-transfer properties, act as light filters, and protect the endoperoxides from being destroyed by the absorption of light.

The tetraphenyl derivative, 1,2,3,4-tetraphenyl-1,3-cyclopentadiene (**25**), is transformed into the corresponding endoperoxide (**26**) via a direct photolysis (*58*). The endoperoxide is easily isomerized to 1,2:3,4-diepoxy-1,2,3,4-tetraphenylcyclopentane (**27**).

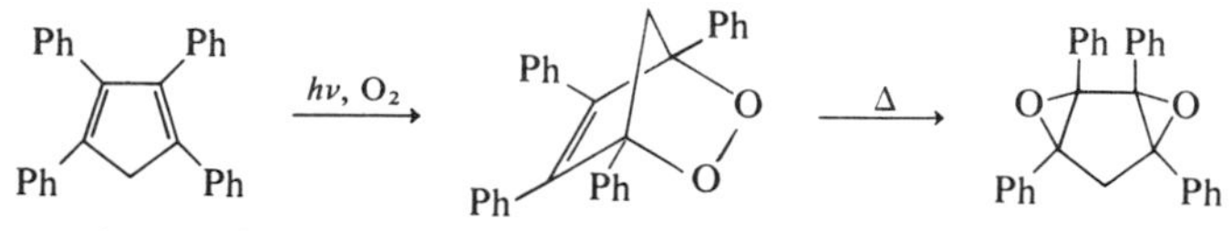

(**25**) $\xrightarrow{h\nu,\ O_2}$ (**26**) $\xrightarrow{\Delta}$ (**27**)

A similar rearrangement occurs in the photooxidation of 2,3,4,5-tetraphenylfulvene (**28**) (*58,59*). In this case the isomerization is so facile that all attempts to isolate the corresponding endoperoxide (**29**) failed.

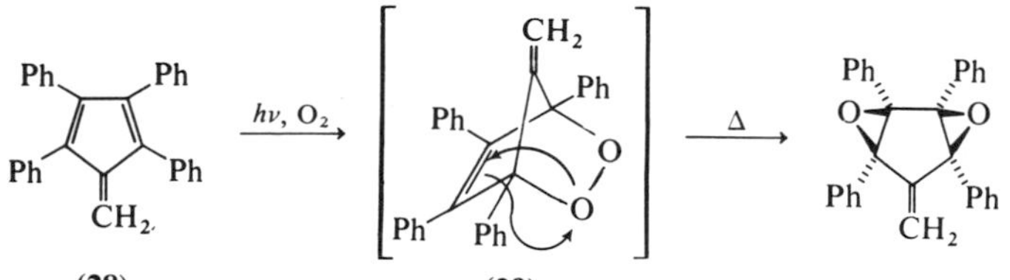

(**28**) $\xrightarrow{h\nu,\ O_2}$ [(**29**)] $\xrightarrow{\Delta}$

Irradiation of a series of benztropolones (**30**, R = H, OH, and OMe) under oxygen in the presence of sensitizer provided high yields of the lactones (**31**, R = H, OH, and OMe) (*60,61*). The intermediate endoperoxide (**32**) decomposed *in situ* by either

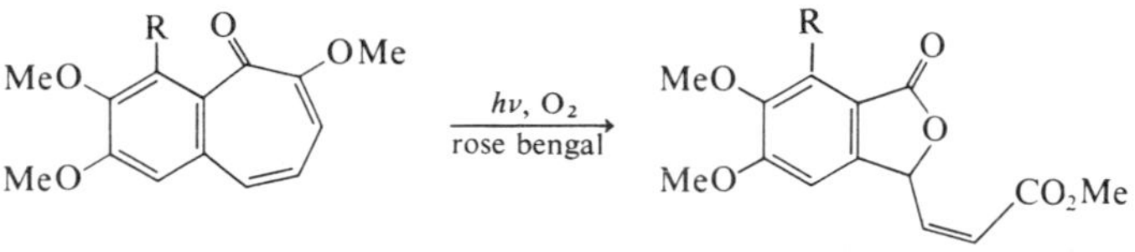

(30) $\xrightarrow[\text{rose bengal}]{h\nu,\ O_2}$ (31)

photolytically or thermally induced routes to the lactone (**31**).

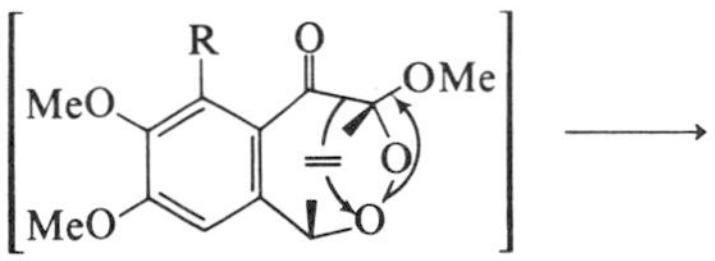

(32) ⟶ (31)

Rio and Berthelot (*62*) have reported the photooxygenation of an acyclic diene. The methylene blue-catalyzed oxidation of *trans-trans*-1,4-diphenyl-1,3-butadiene (**33**) afforded *cis*-3,6-diphenyl-3,6-dihydro-1,2-dioxine (**34**) in excellent yield.

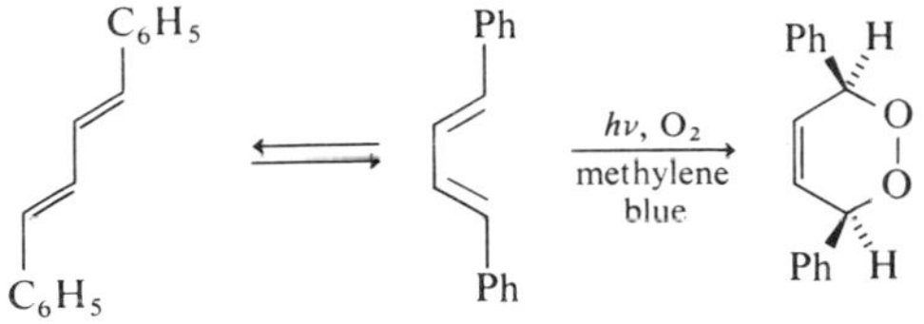

(33) ⇌ $\xrightarrow[\text{methylene blue}]{h\nu,\ O_2}$ (34) 68%

Other conjugated dienes have been subjected to dye-sensitized photooxygenations with comparable results (*13,63–66*).

V. PHOTOOXYGENATION OF HETEROCYCLIC PENTADIENE DERIVATIVES

A. Furans

The dye-sensitized oxidation of furans also proceeds via a cyclic peroxide (a cyclobutadiene monozonide) in a manner analogous to reactions of cyclic dienes. These transannular peroxides are extremely unstable and have been isolated in only a few cases.

Furan itself yields a peroxide (**35**) that explodes at −10°. The peroxide can be isolated from oxidation reactions carried out at −100°. Polymerization of the transannular peroxide occurs at temperatures above −80°, while rearrangement is observed in methanol at room temperature to afford a lactone (**36**) as the only isolable product (*47,67,68*).

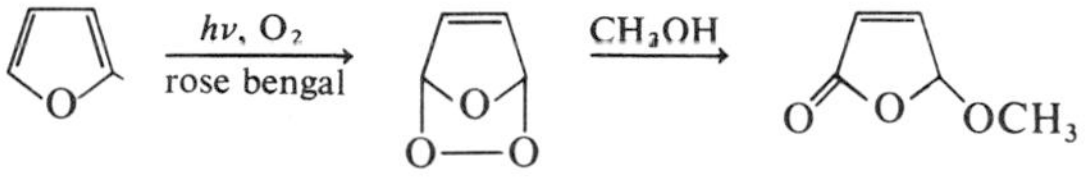

$h\nu$, O_2, rose bengal → (**35**) → CH_3OH → (**36**) 88%

Photooxidation of 2-methylfuran in methanol at room temperature gives rise to the methoxyhydroperoxide (**37**) in high yield (*47,69*).

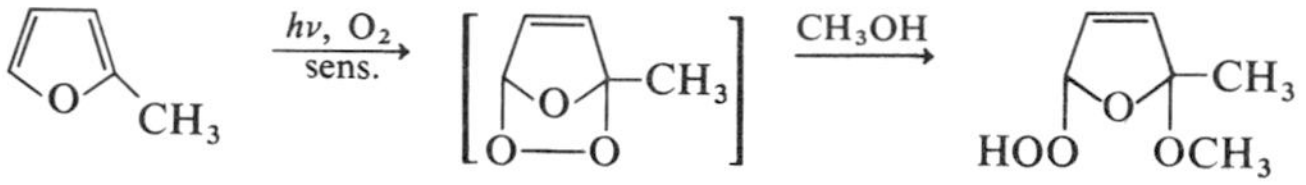

$h\nu$, O_2, sens. → [ ] → CH_3OH → (**37**)

Similar results are obtained with 2,5-dimethylfuran (**38**) (*47,69*).

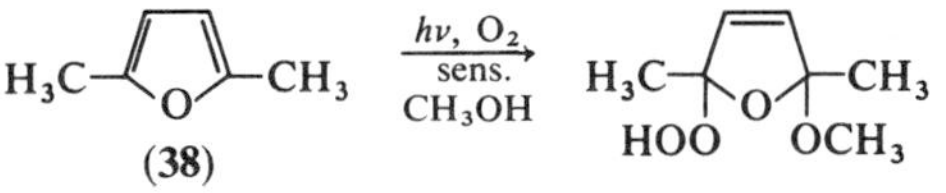

(**38**) → $h\nu$, O_2, sens., CH_3OH →

Diphenylisobenzofuran yields a crystalline peroxide (**39**) that was observed to be stable for several hours at −78°, but that immediately exploded when warmed to 18°. In solution (at ambient temperatures) it is transformed into (**40**) (*70*). The mechanism of the decomposition reaction has not been established, and the fate of the third oxygen atom in (**39**) remains unknown.

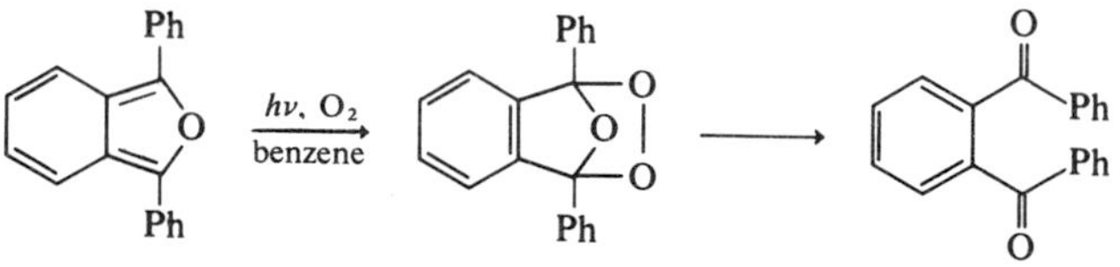

$h\nu$, O_2, benzene → (**39**) → (**40**)

Irradiation of furfural at room temperature in ethanol affords 5-ethoxy-2,5-dihydrofuran-2-one (**41**) (*47,69*). The ether component of (**41**) depends on the alcoholic solvent employed.

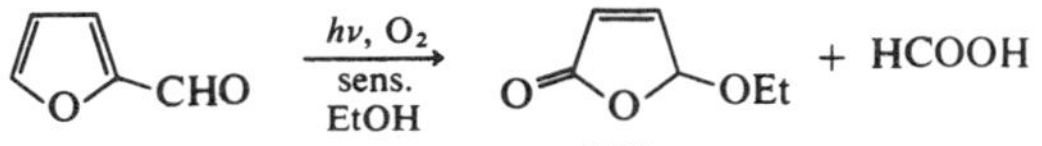

(41) 86%

As with other furans, the transannular peroxide (**42**) is presumed to be the primary product, which subsequently reacts with the alcohol.

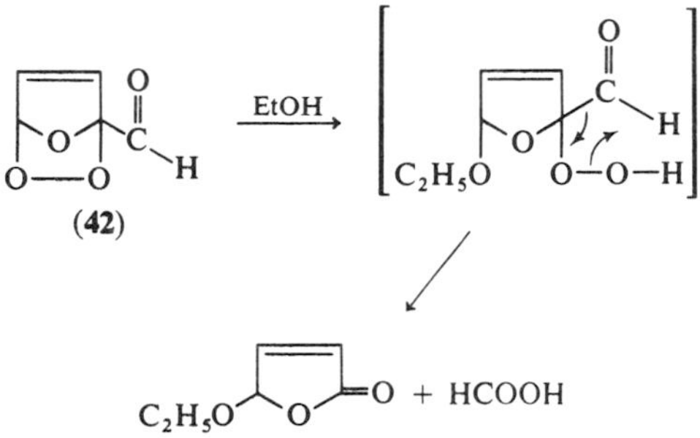

(**42**)

Irradiation of heterocyclophane (**43**) in an aerated methanoic solution ultimately yields a tetracyclic diketone (**46**) (*71*). The reaction appears to occur through an intramolecular Diels–Alder condensation of intermediate (**44**) to yield the hydroperoxidic intermediate (**45**), which has been isolated by Katz and coworkers (*72*).

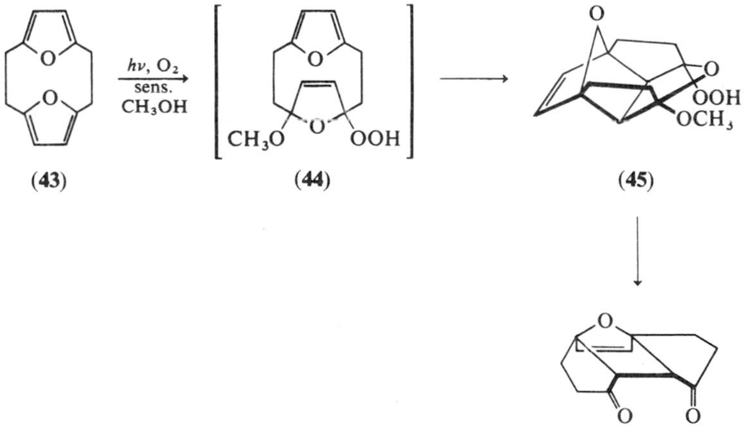

(**43**) (**44**) (**45**)

(**46**) 42%

The photooxygenation of (**43**) takes an entirely different course in methylene chloride, in which the product consists solely of the tetraepoxide (**47**) (*73*).

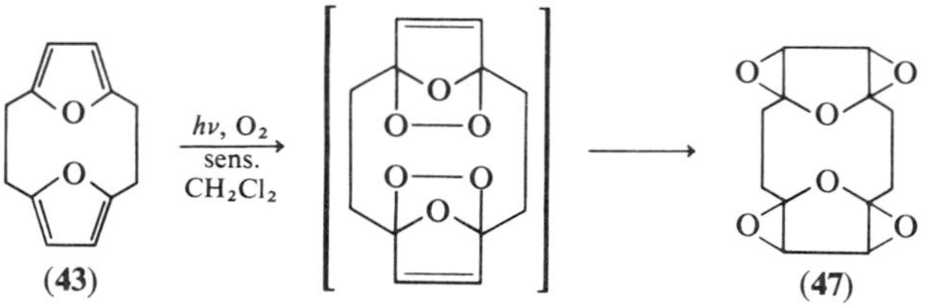
hν, O₂
sens.
CH₂Cl₂

(43) (47)

Similar results have been obtained on photooxidation of other substituted furans (*74–80*). The modes of decomposition of the intermediate peroxides parallel the examples above.

B. Pyrroles

Early work on the photooxidation of pyrroles led, for the most part, to the formation of tars. Bernheim and Morgan (*81*) reported the oxidation of pyrrole to give an unidentified crystalline compound that was later shown to be (**48**) (*82*).

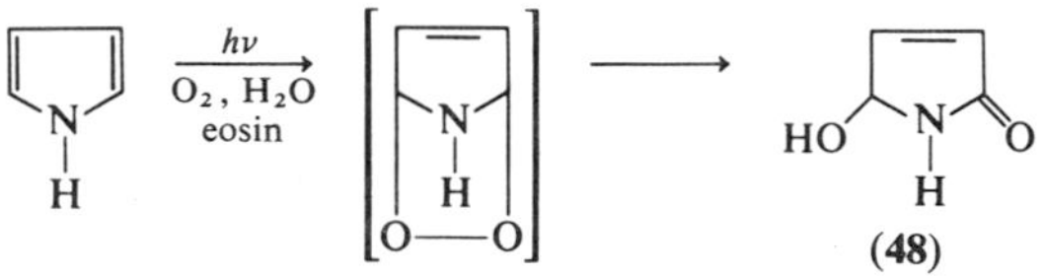
hν
O₂, H₂O
eosin
HO
O

(**48**)
32%

Although the isolation of the transannular peroxide was not attempted, its intermediacy is comprehensible in view of analogous photooxidations of furans.

Similarly, oxidation of *N*-methylpyrrole gave (**49**) in 48% yield.

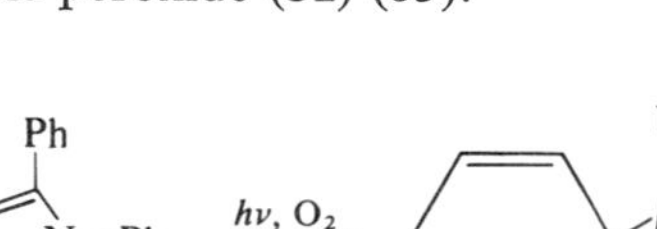
hν, O₂
eosin
H₂O
HO
O

(**49**)

Photooxygenation of 1,2,3-triphenylisoindole (**50**) in carbon disulfide yields a relatively stable peroxide (**51**) (*83*).

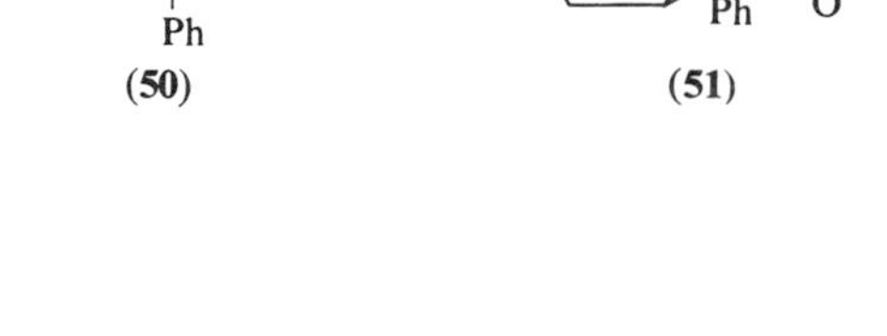
Ph
N—Ph
Ph
hν, O₂
CS₂
Ph
Ph
Ph

(**50**) (**51**)

Irradiation of 9H-pyrrolo[1,2-a]indole (**52**) in a pyridine–tetrahydrofuran–water solvent gives rise to 3H-pyrrolo[1,2-a]indole-3-one (**54**) in 71% yield (*84*). The reaction presumably proceeds via an endoperoxide intermediate (**53**). 2,3,4,5-Tetraphenylpyrrole (**55**) undergoes photooxidation in the presence of methylene blue to yield an epoxide (**56**) and α-*N*-

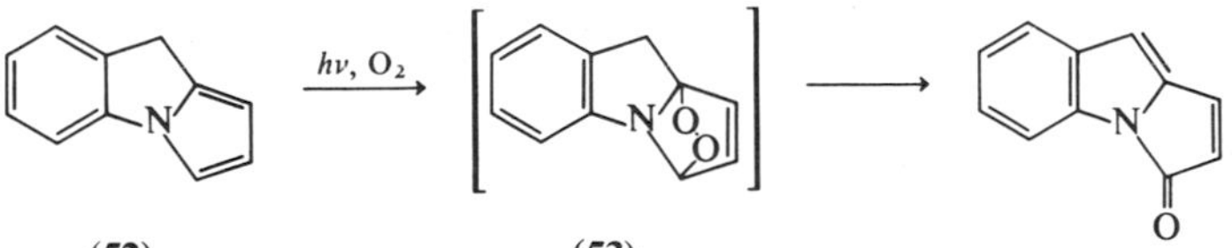

$h\nu$, O_2

(**52**) (**53**) (**54**)

benzoylamino-β-benzoyl stilbene (**57**) (*77*). The formation of the photo-

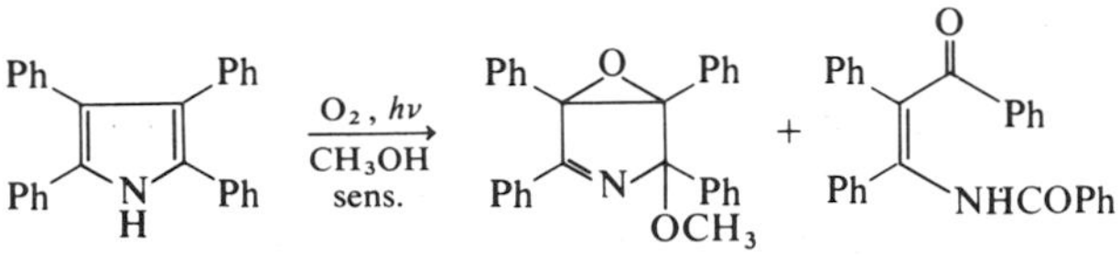

O_2, $h\nu$, CH_3OH, sens.

(**55**) (**56**) 55% (**57**) 30%

oxidation products is explained (*85*) by an isomerization of the transannular peroxide to afford a hydroperoxide, which undergoes further reaction to ultimately yield (**56**) and (**57**). Irradiation of (**55**) in ether, chloroform, or carbon disulfide at −50° gives the hydroperoxide (**58**) exclusively (*85*).

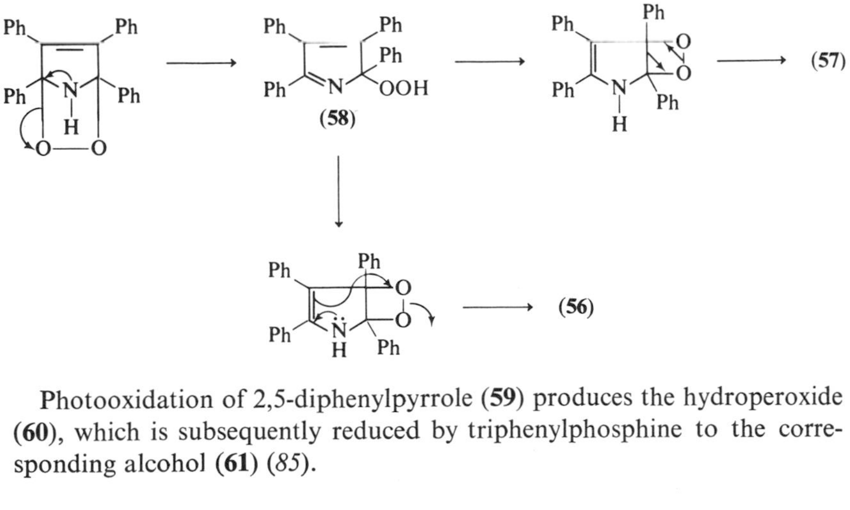

(**58**) → (**57**)

→ (**56**)

Photooxidation of 2,5-diphenylpyrrole (**59**) produces the hydroperoxide (**60**), which is subsequently reduced by triphenylphosphine to the corresponding alcohol (**61**) (*85*).

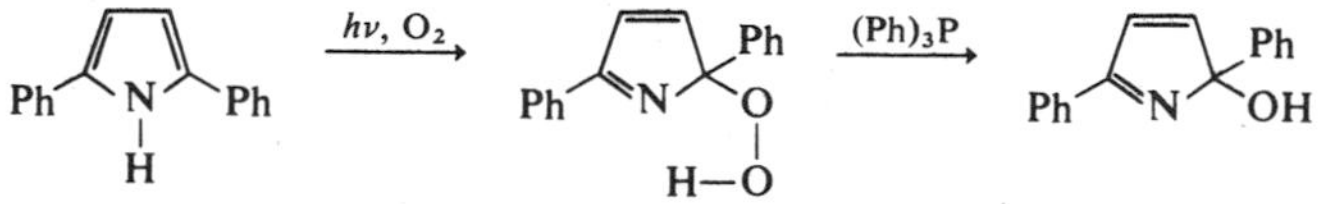

$h\nu$, O_2 ; $(Ph)_3P$

(59) (60) (61)

Prolonged irradiation of 1-methyl-2,3,5-triphenylpyrrole (**62**) at high dilution yielded benzoic acid and *cis*-dibenzoylstyrene oxide (**63**) (*86*).

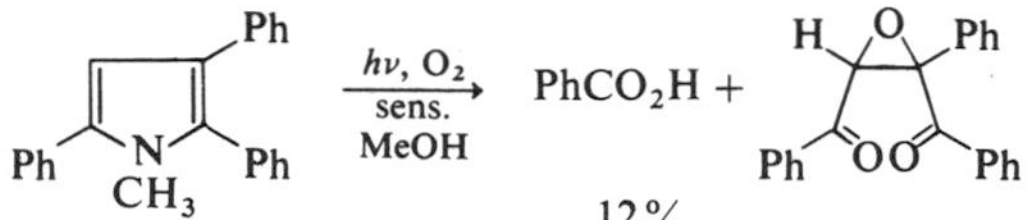

$h\nu$, O_2, sens., MeOH → $PhCO_2H$ + (63)

(62) 12% (63) 65%

C. Oxazoles

The photooxygenation of a variety of oxazoles has been reported. For example, a methanol solution of 2-methyl-5-phenyloxazole (**64**) was converted to benzoic acid and α-acetamido-α-methoxyacetophenone (**65**) (*87*). As in the case of furans, the ozonide-like peroxide (**66**) is believed to be the intermediate. Conversion of (**66**) to the hydroperoxide, followed by methanolysis and subsequent hydrolysis, would lead to the hemiketal (**67**) (*87*).

$h\nu$, O_2, MeOH, sens. → $PhCO_2H$ + $CH_3O(PhCO)CH{-}NH{-}COCH_3$

(64) 83% 10% (65)

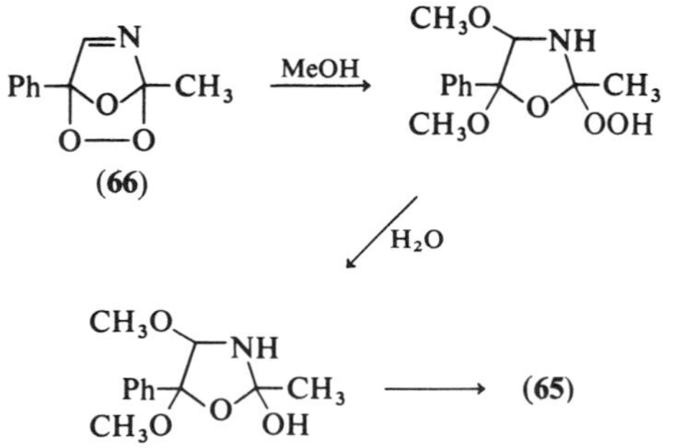

(66) —MeOH→ ; H_2O

(67) → (65)

A different mode of reaction was observed when 2,4,5-triphenyloxazole (**68**) was photooxygenated in methanol or chloroform (*87*).

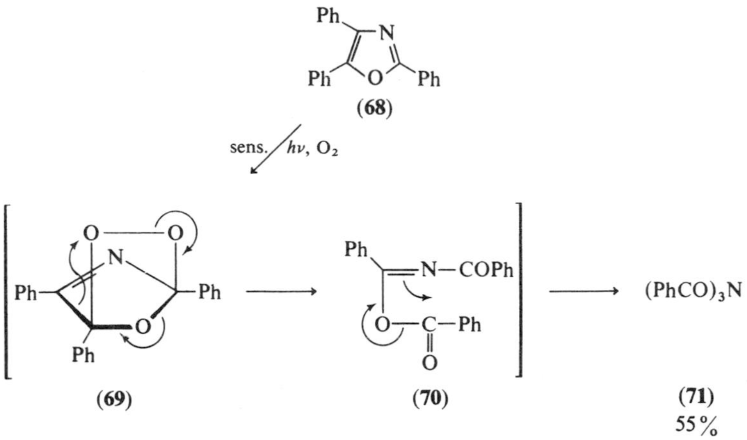

Ph, N, Ph, O, Ph (**68**)

sens./$h\nu$, O_2

[Ph, O—O, N, O, Ph, Ph (**69**) → Ph, N—COPh, O—C(=O)—Ph (**70**)] → $(PhCO)_3N$

(**69**) (**70**) (**71**) 55%

The transannular peroxide (**69**) was presumed to undergo rearrangement to form the *N*-benzoylisoimide (**70**), which then underwent subsequent reorganization to the triamide (**71**). Benzoic acid and benzamide were also isolated. Kurtz and Shechter (*88*) reported the oxidation of (**68**) in ether or benzene to yield tribenzoylamide (**71**), benzonitrile, and benzoic anhydride.

Irradiation of triphenylisooxazole (**72**) in ether or benzene in the presence of oxygen provided tribenzoylamide, phthalic anhydride, benzonitrile, and a ketenimine (**73**) (*88*).

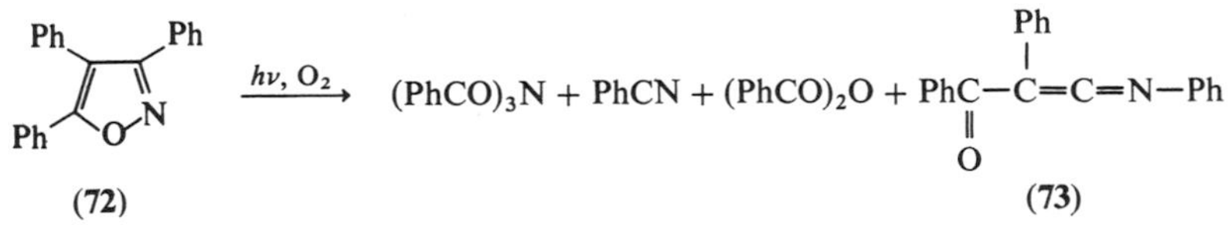

$$\text{(72)} \xrightarrow{h\nu,\ O_2} (PhCO)_3N + PhCN + (PhCO)_2O + PhC(=O){-}C(Ph){=}C{=}N{-}Ph$$

(**72**) (**73**)

The generality of triamide formation in the photooxidation of oxazoles was further shown in studies with 2-methyl-4,5-diphenyloxazole (**74**) and 4-methyl-2,5-diphenyloxazole (**75**) (*87*).

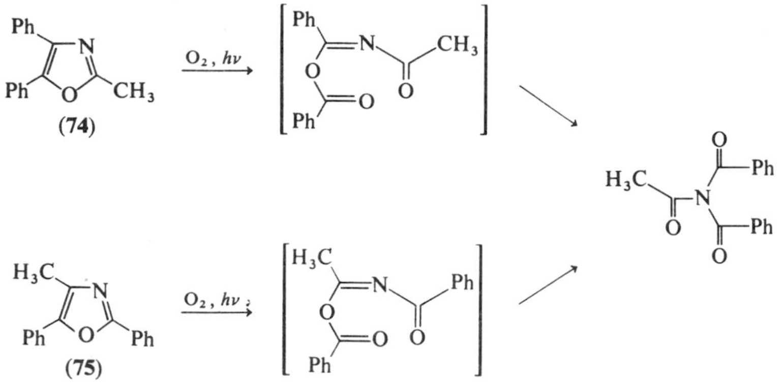

An entirely different course of reaction was observed when 4,5-condensed oxazoles, such as (**76**), were photooxygenated in inert solvents (*89*).

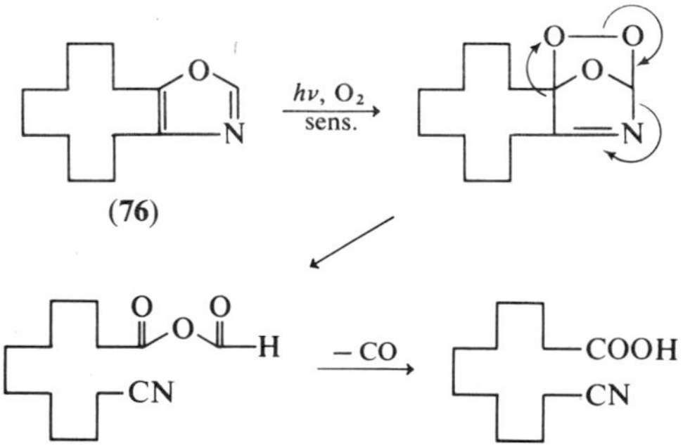

D. Imidazoles

The first example of an imidazole reacting with singlet oxygen was the photooxidation of lophine (2,4,5-triphenylimidazole) (**77**), which yielded a compound that was assumed to be the endoperoxide (**78**) (*90*). Later it was shown that a hydroperoxide (**79**) rather than a peroxide was indeed the reaction product (*91,92*). The parent heterocyclic system, imidazole,

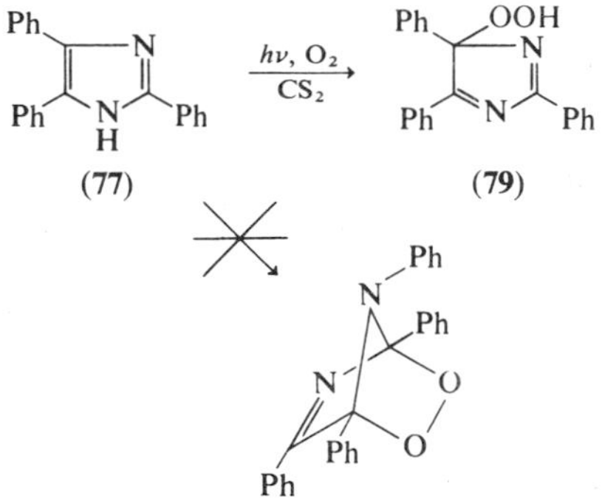

(77) (79) (78)

was observed to undergo a slow oxidation in methanol to give **(81)** (*93*) via the sequence indicated below.

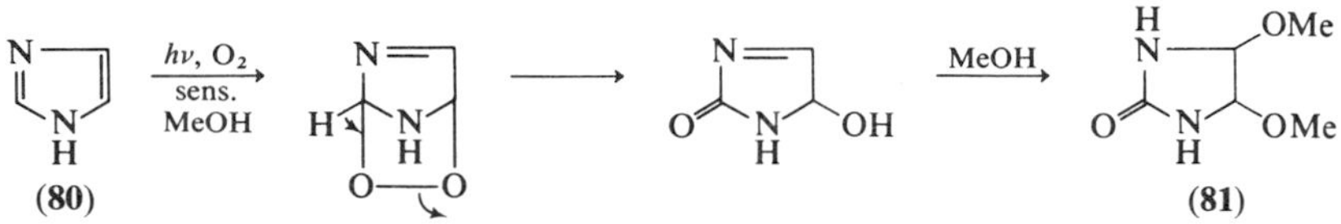

(80) (81)

Tetraphenylimidazole **(82)** underwent photooxidation to yield *N,N′*-dibenzoyl-*N*-phenylbenzamidine **(83)** as the sole product in an example of a peroxide rearrangement

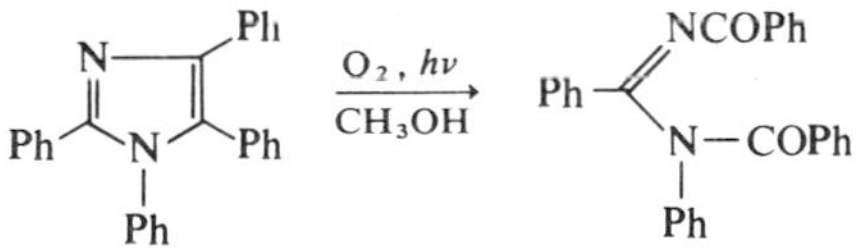

(82) (83)

involving the dioxetane intermediate **(85)** (*93*).

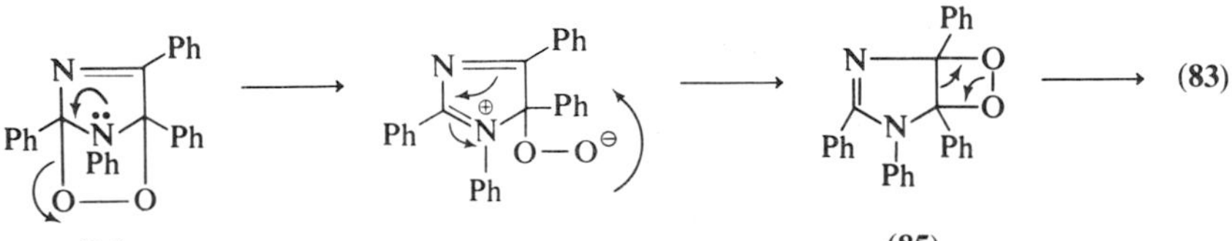

(84) (85)

A methanolic solution of 9-phenylxanthine (**86**) in the presence of rose bengal was reportedly photooxidized to (**87**) in 58% yield (*94*).

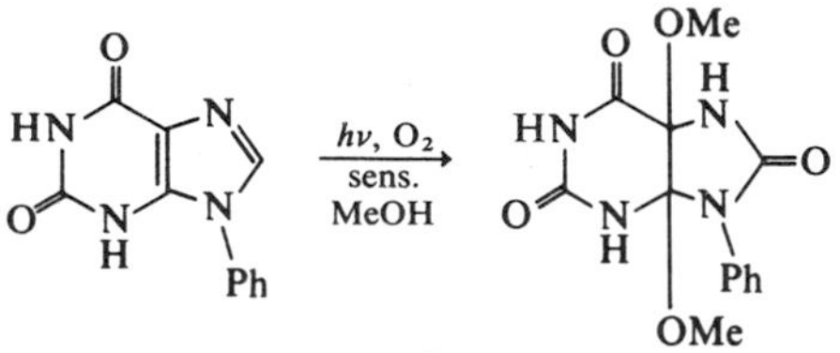

O HN N O N H N Ph — $h\nu$, O_2, sens., MeOH → O OMe H HN N =O O N H N Ph OMe

(**86**) (**87**)

The product formation was rationalized on the following basis. A conventional transannular peroxide was solvolyzed by methanol to a methoxyhydroperoxide (**88**), which was subsequently dehydrated to give an 8-keto compound (**89**). Addition of a second molecule of methanol to the imine double bond gave rise to the observed product.

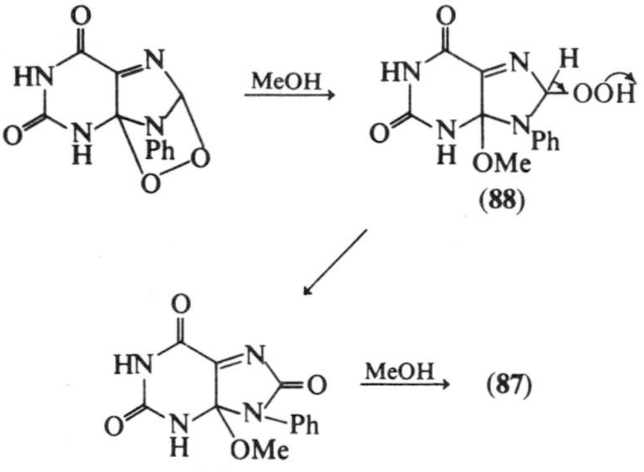

O HN N O N H N Ph O O — MeOH → O HN N H OOH O N H N Ph OMe

(**88**)

O HN N =O O N H N Ph OMe — MeOH → (**87**)

(**89**)

Irradiation of 8-methoxycaffeine (**90**) in the presence of rose bengal and air provided 1-methyl-2,2-dimethoxy-4-methylamino-3-imidazolin-5-one (**91**) in 78% yield (*95*).

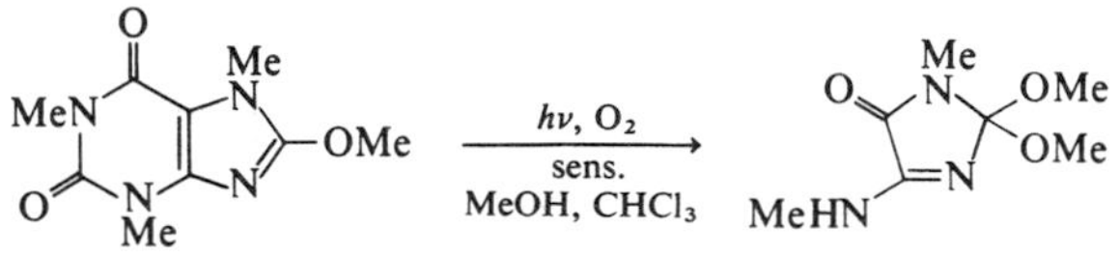

O Me MeN N —OMe O N Me N — $h\nu$, O_2, sens., MeOH, $CHCl_3$ → Me O N OMe OMe N MeHN

(**90**) (**91**)

The formation of imidazolinones by this procedure has been explained by the series of transformations shown in Scheme 6. The mechanism was supported by a trace experiment employing (8-^{14}C)-labeled 8-methoxycaffeine (*95*).

Scheme 6

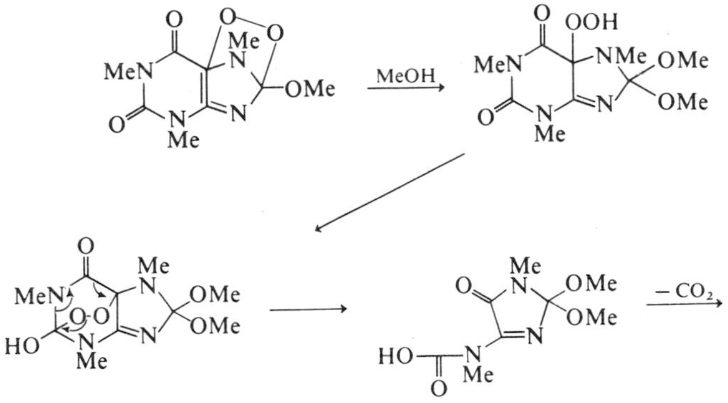

Comparable results have been reported for related photooxidations (*93,96*).

E. Thiophenes

Unlike their oxygen and nitrogen analogs, thiophene and tetraphenylthiophene were found to be unreactive to singlet oxygen (*75,97*). However, alkylsubstituted thiophenes such as 2,5-dimethylthiophene (**92**) have been reported to undergo a facile reaction with singlet oxygen (*98,99*).

Irradiation of (**92**) in chloroform with methylene blue afforded the *cis*-sulfine (**93**) and the diketone (**94**).

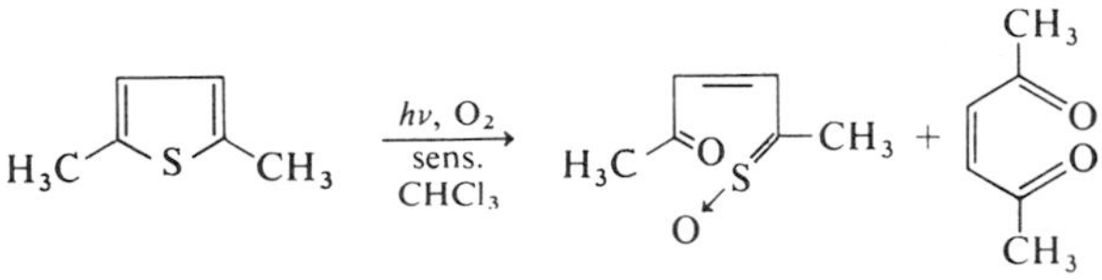

(**92**) (**93**) 56% (**94**) 28%

Oxidations carried out in methanol gave the sulfine (**93**) and the diketone (**94**) in 70 and 2% yields, respectively.

The formation of product is presumed to occur via a thioozonide intermediate (**95**), analogous to other heterocyclic systems, or by a sulfoxonium intermediate (**96**) formed by the attack of singlet oxygen on the sulfur atom of the thiophene (*98,99*).

(**92**) $h\nu$, O_2 sens. (**95**) $-S$ → (**94**); (**96**) → (**93**)

Photooxidation of 2,3-dimethyl-4,5,6,7-tetrahydrothionaphthene (**97**) in methanol gave the sulfoxide (**99**) in 10% yield (*98*). The formation of (**99**) occurs by the addition of solvent to the intermediate sulfine (**98**).

(**97**) $\xrightarrow[\text{sens. } CH_3OH]{h\nu,\ O_2}$ (**98**) → (**99**)

VI. PHOTOOXYGENATION OF ACENES

Polyaromatic systems such as naphthacene, pentacene, hexacene, 1-aza-, and 2-azaanthracene undergo facile photooxygenation to yield transannular peroxides, but no oxidation is observed with phenanthrene

(*100*), naphthalene (*101*), or acridine (*101*). Typical of this class of reaction is the photooxidation of anthracene to give the 9,10-transannular peroxide (**100**) in quantitative yield (*102*).

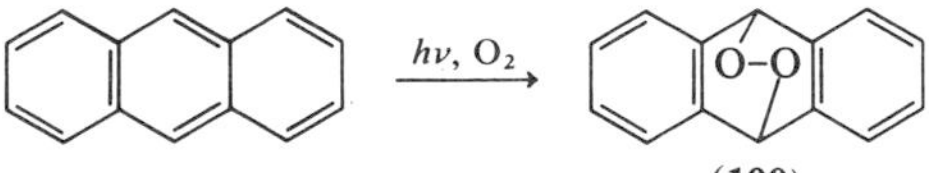

$h\nu$, O_2

(**100**)

More than one hundred photoperoxides of the anthracene and naphthacene series have been synthesized, and they as well as other aromatic hydrocarbons have been the subject of several reviews (*13,14,20*). This section will be restricted to the more recent publications in this area.

While naphthalene itself is unreactive to singlet oxygen, direct photooxygenation of the strained 1,4-dimethoxy-5,8-diphenylnaphthalene (**101**) at $-50°$ results in the formation of the photooxide (**102**) in 80 % yield (*103*).

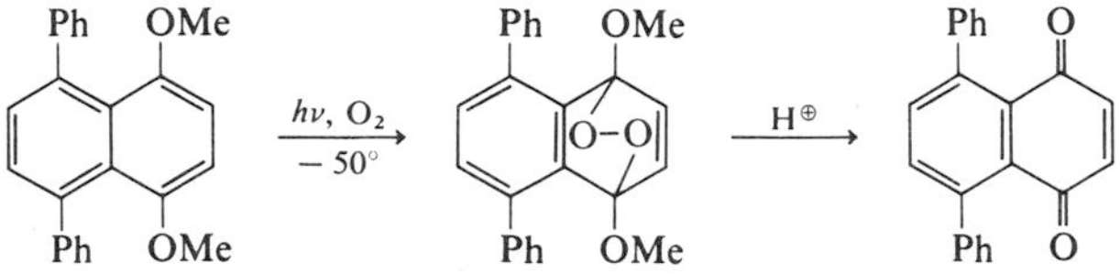

$h\nu$, O_2, $-50°$; $H^{\oplus}$

(**101**) (**102**)

The anti[2.2]paracyclonaphthane (**103**) reacts with singlet oxygen to form a transannular peroxide (**104**) which undergoes a Diels–Alder condensation to yield (**105**). Finally, (**105**) is transformed by solvolysis in methanol to a novel polyclic product (**106**) (*104*).

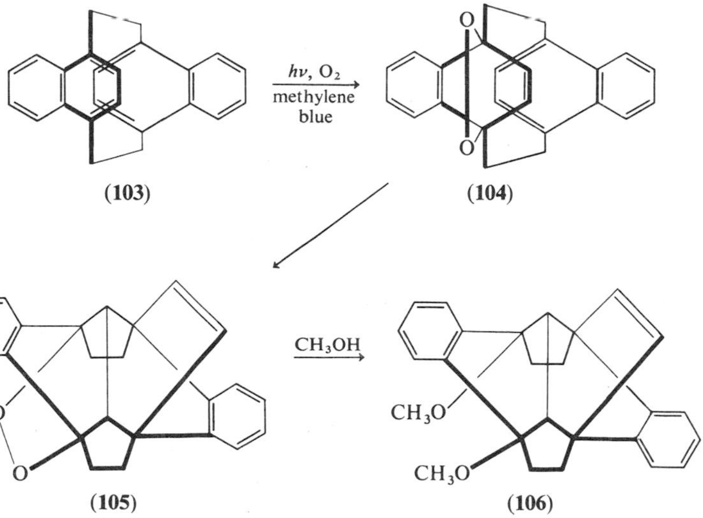

$h\nu$, O_2, methylene blue

(**103**) (**104**)

CH_3OH

(**105**) (**106**)

VII. PHOTOOXYGENATION OF OLEFINS

A. Monoenes and Dienes

Dye-sensitized photooxidation of olefins containing at least one allylic hydrogen provides a useful way of introducing an allylic oxygen function with a concomitant rearrangement of the double bond. In keeping with the electrophilic character of singlet oxygen, a tetrasubstituted double bond may be attacked preferentially to a trisubstituted double bond in the same molecule. The observed order of reactivity is tetra- > tri- ≫ di-, while monosubstituted double bonds are essentially unreactive. Due to the selective reactivity of singlet oxygen, a large excess of oxygen or a high concentration of substrate is necessary if a useful conversion of a slightly reactive compound is desired.

Studies with cyclic olefins and steroidal systems indicate that the reaction proceeds by a cyclic abstraction mechanism with rather stringent geometric requirements (*105–107*). These rigid systems require that a *cis* relationship exist between the carbon–hydrogen bond cleaved and the carbon–oxygen bond formed. Studies have also revealed that pseudoaxial allylic hydrogens are removed in preference to pseudoequatorial hydrogens in steroidal olefins (*3*) and that electron-withdrawing substituents geminal to allylic hydrogens reduce the rate of reaction with singlet oxygen (*38*).

The dye-sensitized photooxygenation of optically active limonene (**107**, Fig. 2) unequivocally differentiates between the singlet oxygen process and free-radical chain reactions (*107*). Although *cis-* and *trans-*hydroperoxides were formed in both reactions, those obtained from the photosensitized process were optically active, and those obtained from the free-radical autoxidation were racemic. Similar product ratios were obtained from the chemical oxidation (NaOCl, H_2O_2) of (**107**) (*108*).

The stereoselectivity of photooxygenation can be influenced by steric as well as conformational effects. The observed product distribution from the oxidation of Δ^3-carene (**108**) was explained (*109*) by the assumption that an equilibrium existed between conformers (**109**) and (**110**) and that only the closed boat form (**110**) reacted. Thus, the allylic hydrogen atoms at C-2 and C-5 are in an axial position only when they are *cis* to the cyclopropane ring (**109**), but in (**110**) they are *trans* to the cyclopropane ring. Since only the *trans* alcohols are formed, the bulky dimethylcyclopropane ring apparently inhibits a *cis*-oxygen attack.

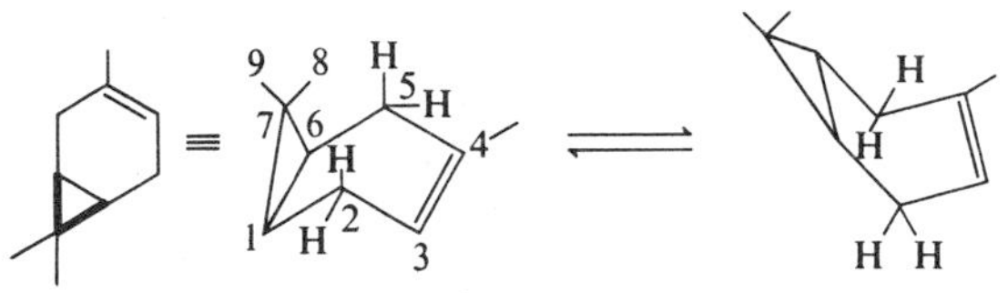

(108) (109) (110)

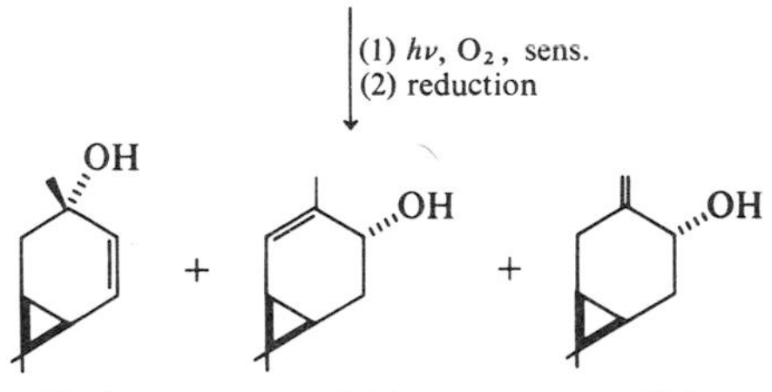

(1) $h\nu$, O_2, sens.
(2) reduction

50% 27% 23%

Bulky groups such as the C-8 methyl group of Δ^2-carenes (**111** and **112**) also shield the double bond from attack by oxygen (*110*).

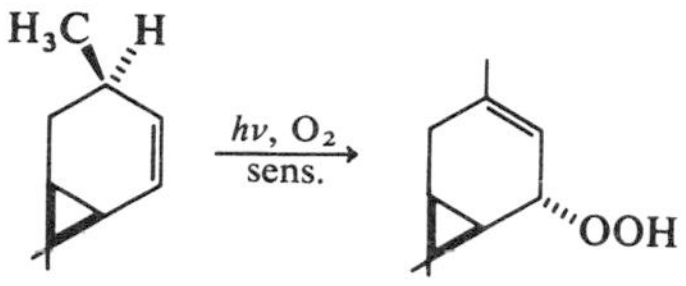

$\xrightarrow[\text{sens.}]{h\nu,\ O_2}$

(111)

$\xrightarrow[\text{sens.}]{h\nu,\ O_2}$ No reaction

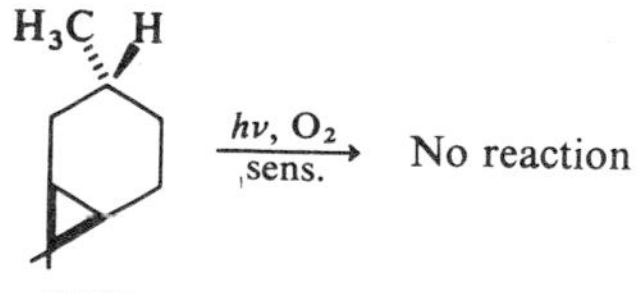

(112)

The remarkable stereoselectivity of the photooxygenation reaction was elegantly shown by the oxidation of chloresterol-7α-d (**113a**) to 3β-hydroxy-5α-hydroperoxy-Δ^6-cholestene (**114a**), the product retaining only 8.5% of the original deuterium (**105**). Similar oxidation of chloresterol-7β-d (**113b**) gave (**114b**) with 95% of the original deuterium retained.

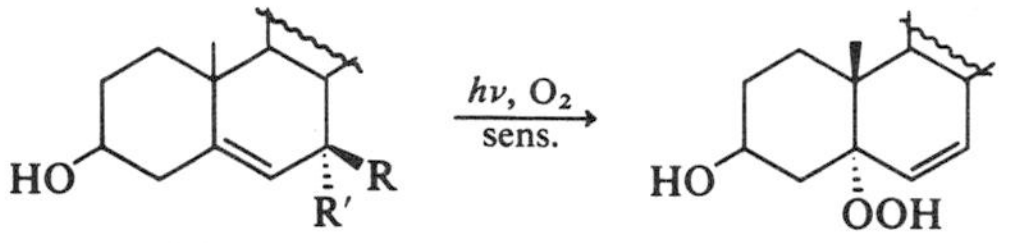

$\xrightarrow[\text{sens.}]{h\nu,\ O_2}$

(113) (114)

(a) R = H, R′ = D
(b) R = D, R′ = H

(1) free radical autoxidation
(2) reduction

HO H 11% + HO H 12% + OH 22% (racemic) + OH 19% + O 13% + O 13% (racemic) + unidentified products 10%

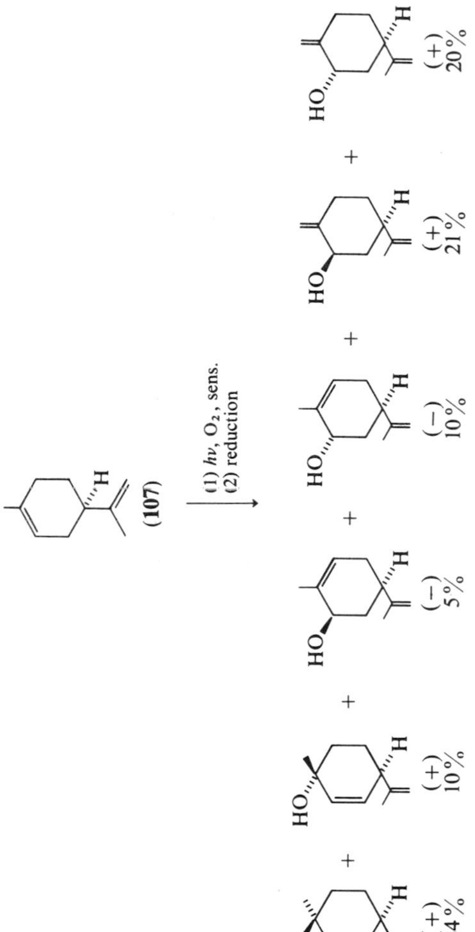

(107)

(1) $h\nu$, O_2, sens.
(2) reduction

(+) 34% + (+) 10% + (−) 5% + (−) 10% + (+) 21% + (+) 20%

Fig. 2. Autoxidation and dye-sensitized photooxidation of (+)-limonene (**107**).

The dye-sensitized photooxygenation of 2-ethylidenebicyclo[2.2.1]hept-5-ene (**115**) carried out at ambient temperatures was found to yield 2-*exo*- and 2-*endo*-hydroxy-2-vinylbicyclo[2.2.1]hept-5-ene (**116**) and (**117**), 2-(α-hydroxyethyl)bicyclo[2.2.1]hepta-2,5-diene (**118**), and tricyclo-[4,3,0,0^{2,7}]non-8-en-4-one (**119**) (*111*).

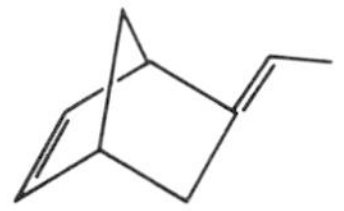

(**115**)

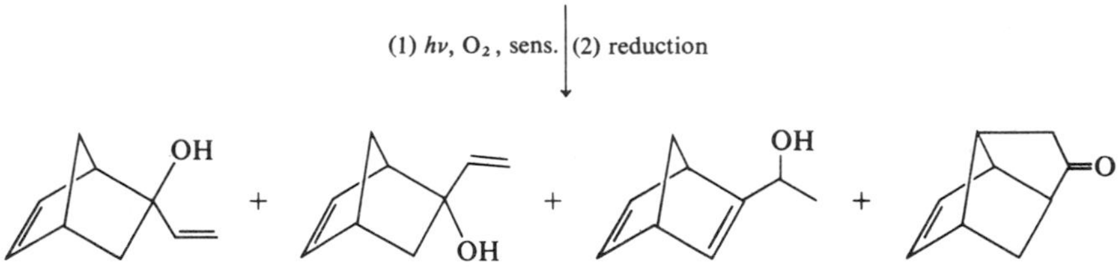

(1) $h\nu$, O_2, sens. (2) reduction

OH + OH + OH + O

(**116**) 35%
(**117**) 12%
(**118**) 23%
(**119**) 5%

The observed predominance of the *exo*-hydroxy epimer (**116**) must be due to selective approach of the singlet oxygen to the more excessible *exo* side of the molecule. For example, the attack of singlet oxygen on 2-methylbicyclo[2.2.1]hept-2-ene (**120**) occurs exclusively from the *exo* direction to afford the *exo* alcohol (**121**) after reduction (*111*).

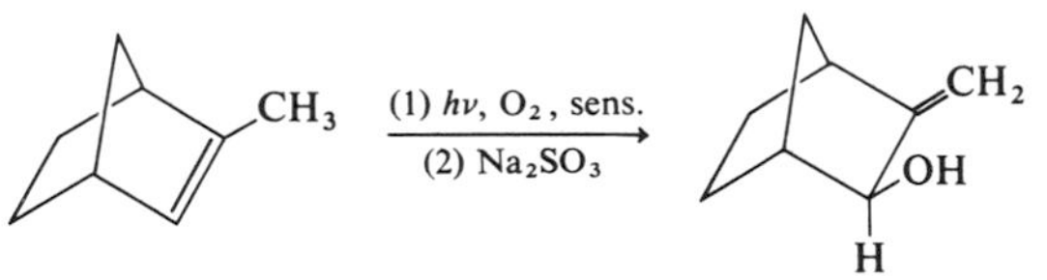

CH_3 (1) $h\nu$, O_2, sens. (2) Na_2SO_3 → CH_2 OH H

(**120**) (**121**)

The production of the novel tricyclic ketone (**119**) appears to be temperature dependent since photolysis at lower temperatures (10–20°) afforded little or no ketone fraction. The formation of the ketone presumably arises from the thermal breakdown of the *exo*- and *endo*-hydroperoxides (**122**) to an unsaturated ketone (**123**), followed by an intramolecular Diels–Alder condensation.

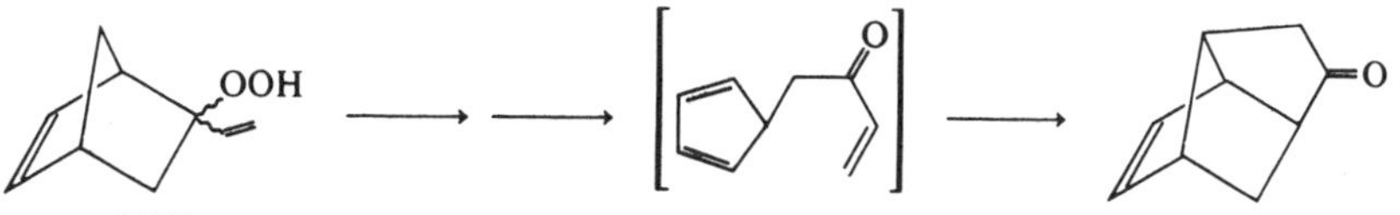

(122) (123) (119)

To account for the conspicuous number of examples in which carbonyl fragments are the major or exclusive products, 1,2-dioxetanes have been proposed as intermediates in the reaction of unsaturated substrates with singlet oxygen. For example, the methylene blue-sensitized photooxidation of indene (**124**) in methylene chloride resulted in the exclusive formation of homophthalaldehyde (**127**), apparently via the 1,2-dioxetane intermediate (**126**) (*2*).

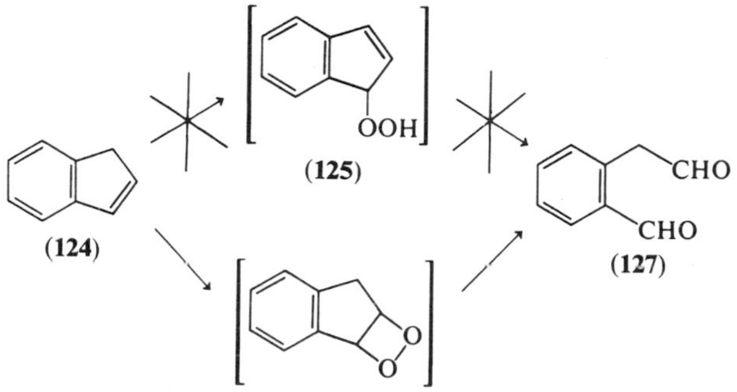

(124) (125) (126) (127)

The intermediacy of (**125**) was ruled out by experiments in which the hydroperoxide was subjected to identical photooxidation conditions. Only 1-indenone was formed (*2*).

1,2-Dioxetanes have also been considered as possible intermediates in the photooxidation of enamines (*112*). For example, a ketone or aldehyde, and an amide were obtained from the enamines (**128a–c**), presumably from the cleavage of an oxygenated intermediate (**130**). The 1,2-dioxetane

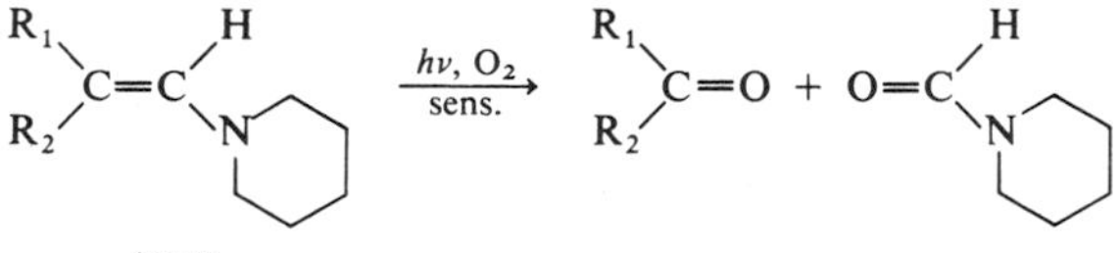

(128)

(**a**) $R_1 = R_2 = CH_3$
(**b**) $R_1 = R_2 = C_6H_5$
(**c**) $R_1 = C_6H_5$, $R_2 = H$

(**129**) was suggested as an intermediate in the formation of (**130**) (*112*).

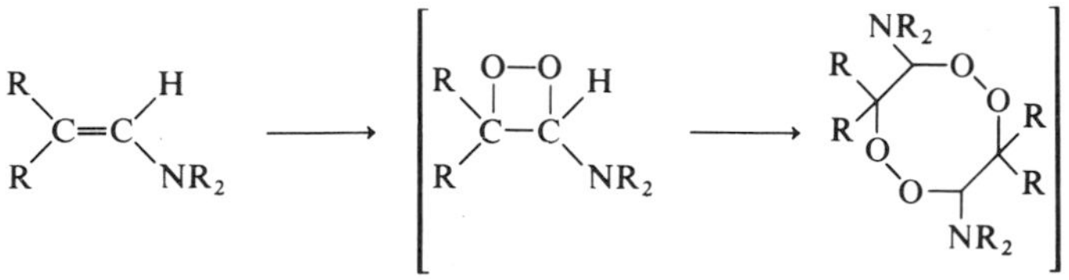

(**129**) (**130**)

Somewhat surprisingly, the enamine (**131**) and β-methoxystyrene (**132**) failed to react under the identical photooxygenation conditions (*112*).

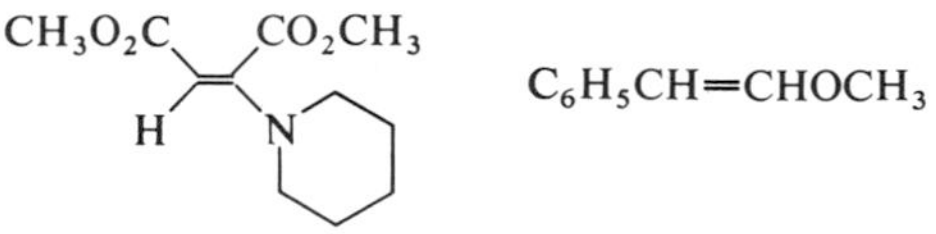

$C_6H_5CH{=}CHOCH_3$

(**131**) (**132**)

B. Polyolefins

Homoannular dienes are more susceptible to photooxidation than are isolated monoolefins. The cyclic diene of ergosterol (**133**) is readily oxidized, but the isolated double bond remains unchanged (*49*).

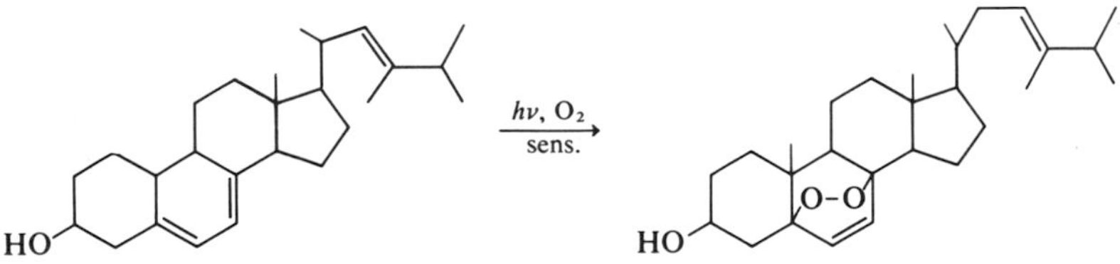

(**133**)

An uptake of two molecules of oxygen occurs with olefins such as isotetralin (**134**) (*13*) and neoabietic acid (**135**) (*113*).

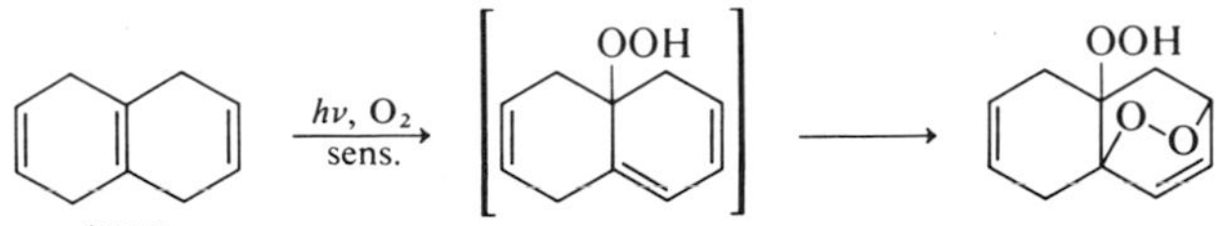

(**134**)

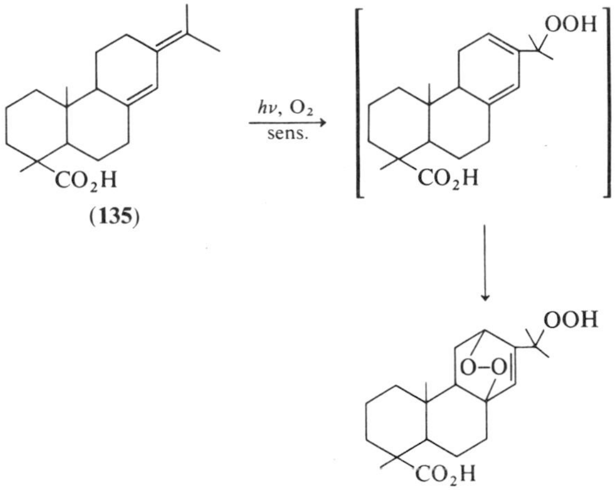

CO_2H
(**135**)

$h\nu$, O_2 / sens.

OOH
CO_2H

OOH
O–O
CO_2H

Photooxygenation of terpinolene (**136**) can be controlled to give two products (**137** and **138**) in nearly equal amounts (*79*).

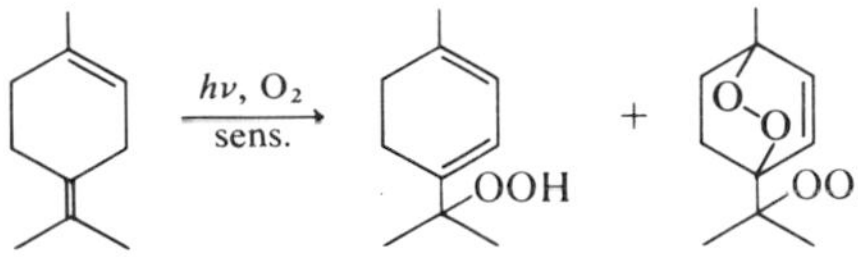

$h\nu$, O_2 / sens.

(**136**) (**137**) OOH + (**138**) OOH

An allenic oxygenation product (**140**), as well as an endoperoxide (**141**) and the alcohols (**142**) and (**143**), are obtained from the photooxygenation of 3-methyl-1-(2,6,6-trimethylcyclohexen-1-yl)-1,3-butadiene (**139**) (*114*).

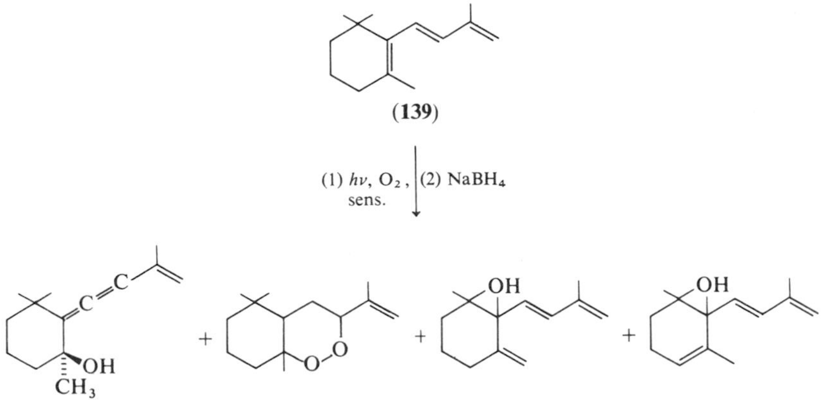

(**139**)

(1) $h\nu$, O_2, sens. (2) $NaBH_4$

(**140**) + (**141**) + (**142**) + (**143**)

Table VI gives representative examples of the photooxidation of various olefins.

TABLE VI

Photooxidation of Olefins

Substrate	Products after reduction	Reference
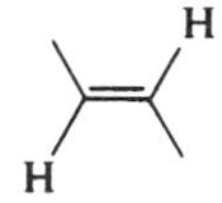 H, H	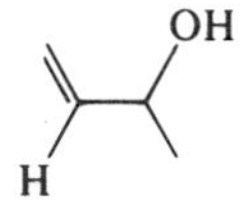 OH, H	
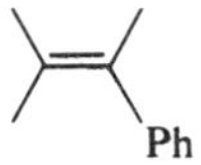 Ph	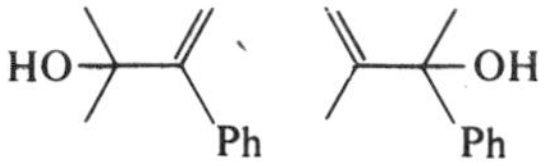 HO, Ph (73%); OH, Ph (27%)	*115* (see also *15,16*)
	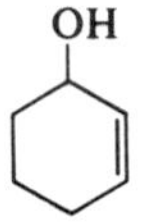 OH	*116* (see also *14,15*)
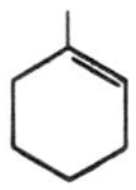	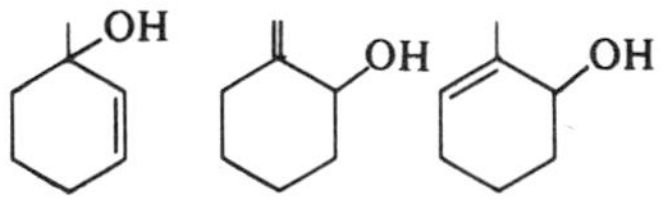 OH (45%); OH (40%); OH (15%)	*117*
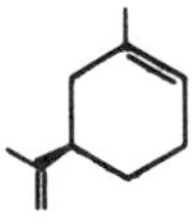	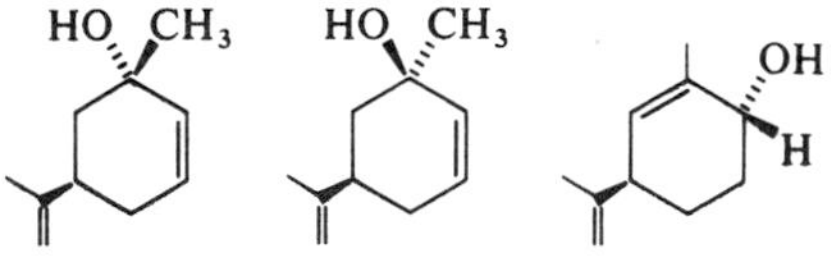 HO, CH_3; HO, CH_3; OH, H	*16,79*
	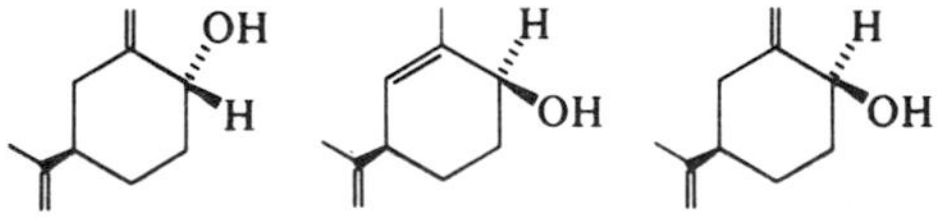 OH, H; H, OH; H, OH	*14*
 H	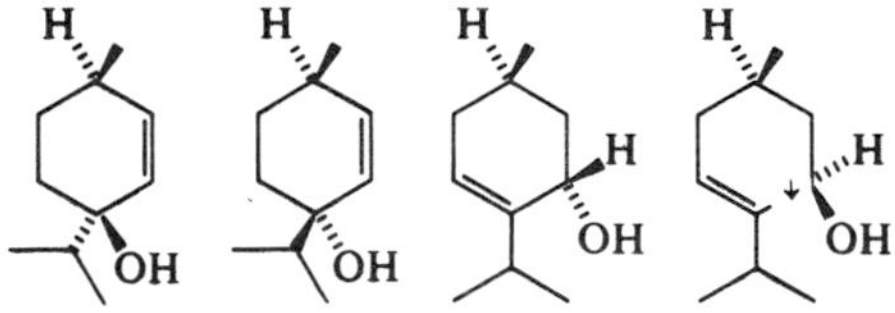 H, OH; H, OH; H, H, OH; H, H, OH	*118*

TABLE VI—*Continued*

Substrate	Products after reduction	Reference
	18% 82%	*110*
	94% 1%	*119*
	80% + 20%	*120*
	88% 12%	*16,79*
		121
	+ + + +	*122*

TABLE VI—*Continued*

Substrate	Products after reduction	Reference
	H, CH_3 ... OH + H, CH_3 ... OH + OH	*123*
	OH + OH + H, CH_3 ... OH	*123*
	CH_2OH	*123*
	CH_2OH	*119* (see also *110,124,125*)
C_8H_{17}, HO	OH	*14*
C_9H_{19}, HO	OH 76% + C_9H_{19} OH 20%	*14*

TABLE VI—*Continued*

Substrate	Products after reduction	Reference
	C_9H_{19}; OH; 4%	*14*
C_8H_{17}; AcO; H; H	+ O + OH; O + O + O	*126*
C_8H_{17}	OH, 29% + H, OH, 37% + HO, H, 7.5%	*127*
	H OH, 8.5%; OH, 2%; O, 6.5%	*105,106*
C_8H_{17}; H; H	H; H; OH	*125*
$=CH_2$	$-CH_2OH$	*125* (see also *128,129*)

TABLE VI—*Continued*

Substrate	Products after reduction	Reference
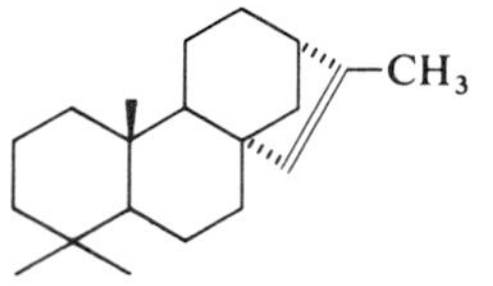 $-CH_3$	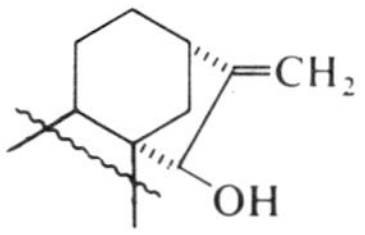 $=CH_2$, OH	*111*
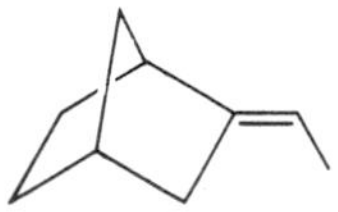	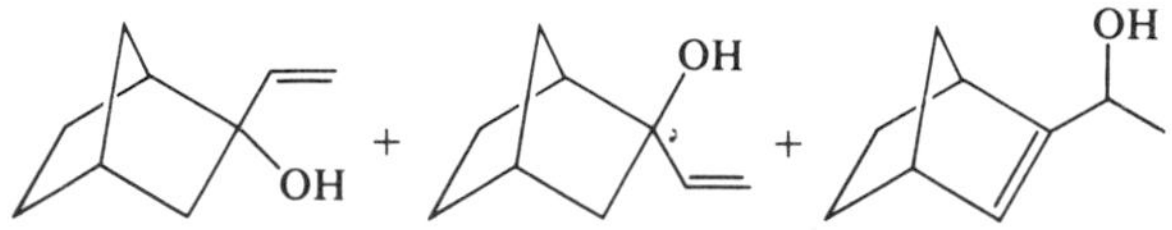 OH + OH + OH	*2*
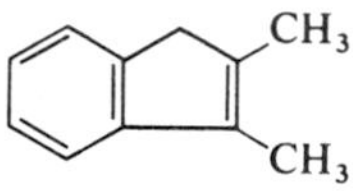 CH_3, CH_3	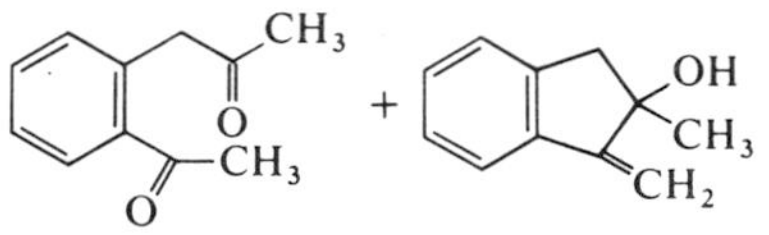 CH_3, O, $-CH_3$, O + OH, CH_3, CH_2	*112*
$CH_3C\equiv C-N(Et)_2$	$CH_3-\overset{O}{\overset{\|}{C}}-\overset{O}{\overset{\|}{C}}-N(Et)_2$	*130*
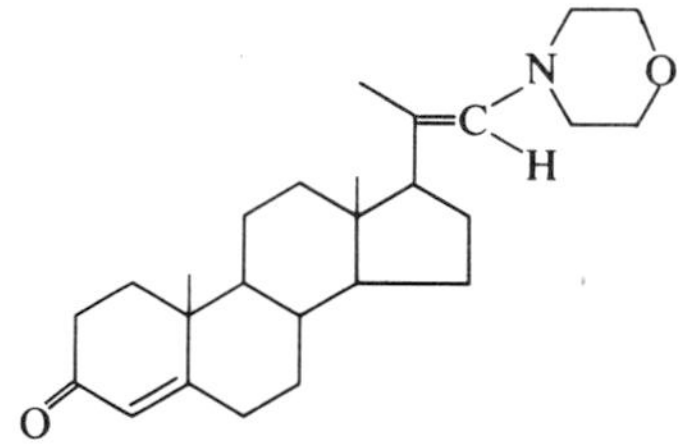 N, O, C, H, O	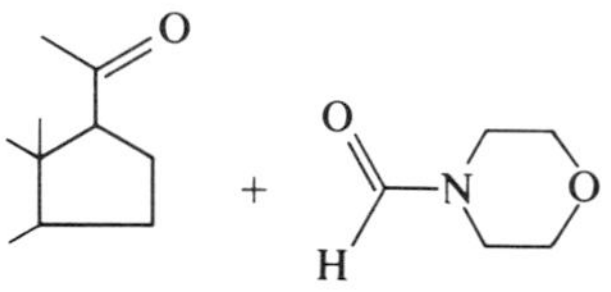 O + O, N, O, H	*14*
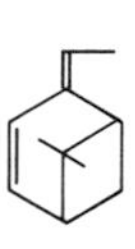	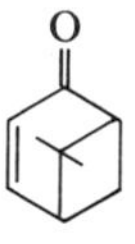 O + CH_3CHO	*14*
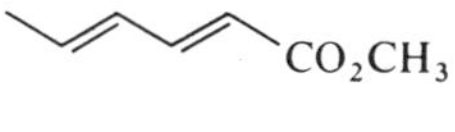 CO_2CH_3	CH_3CHO + OHC CO_2H 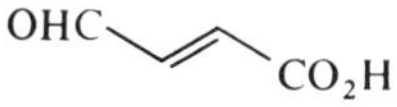	*14*
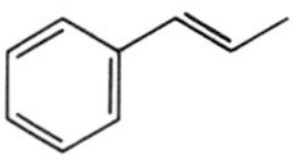	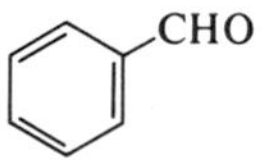 CHO + CH_3CHO	*14*
	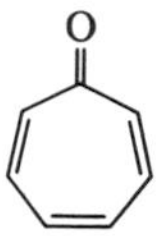 O	*14*

TABLE VI—*Continued*

Substrate	Products after reduction	Reference
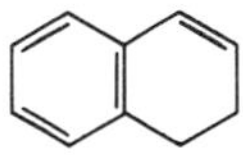	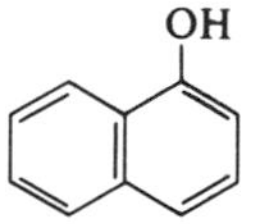 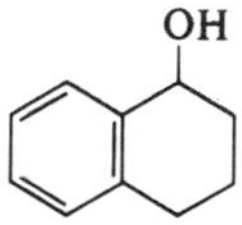 OH + OH	*131* (see also *132,133*)
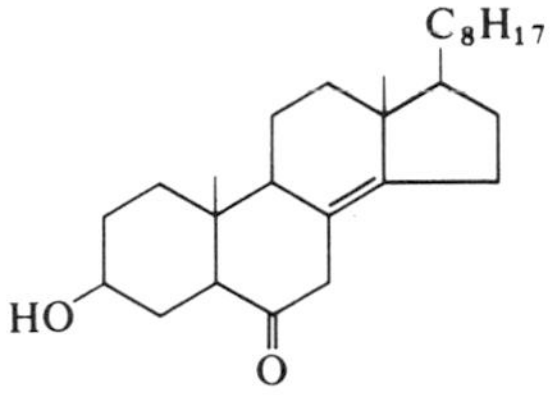 C_8H_{17} HO O	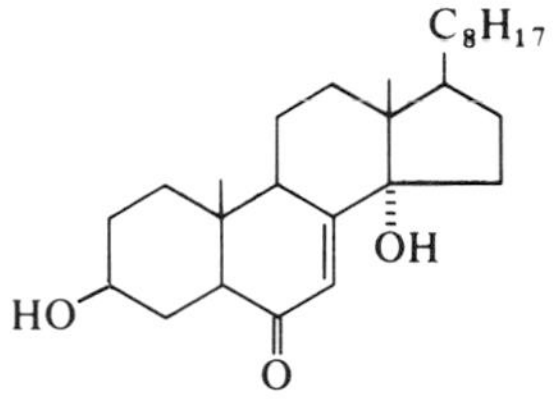 C_8H_{17} HO O OH	*96*
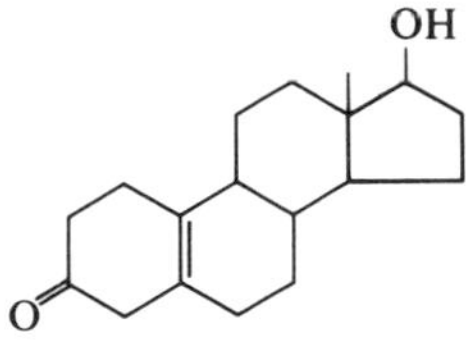 OH O	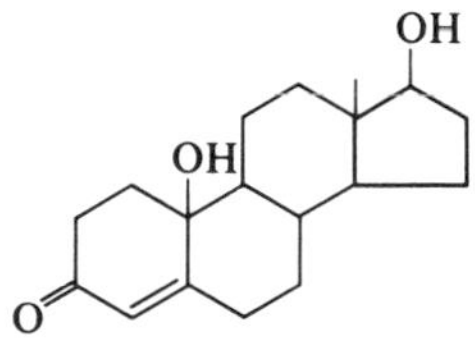 OH OH O	*96*
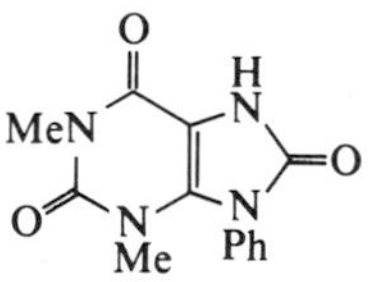 O MeN H N O O N Me N Ph (in MeOH)	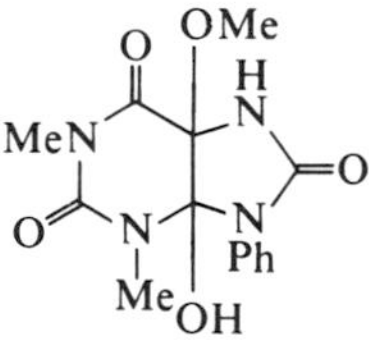 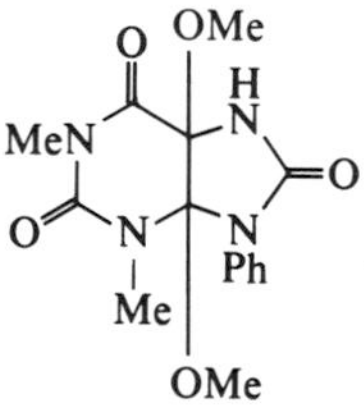 O OMe MeN H N O O N Me N Ph OH 21% ; O OMe MeN H N O O N Me N Ph OMe 2%	*95,134*
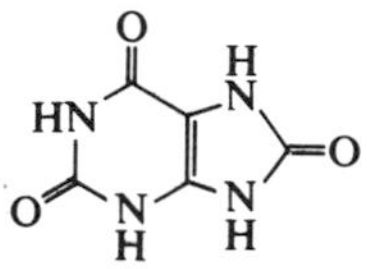 O HN H N O O N H N H in NaOH (aq)	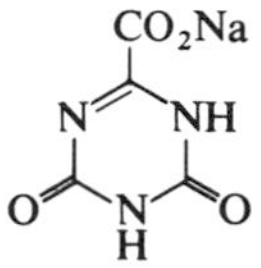 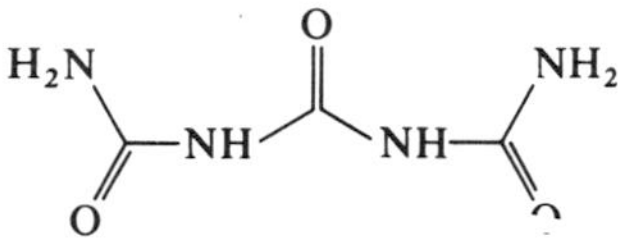 CO_2Na N NH O N H O 40% + O H_2N NH NH NH_2 O O 20%	*134*

TABLE VI—*Continued*

Substrate	Products after reduction		Reference
(in MeOH)	35%	5%	*135*
			135
(in $CHCl_3$)	42%	7%	*66*
			136 (see also *137*)

TABLE VI—*Continued*

Substrate	Products after reduction	Reference
R; R = CH_3 =$CHCO_2CH_3$	R, O–O, 60%; CH_2R, 10%; + R, O–O, 6%	*138* (see also *139*)
OH, CH_3, in NaOH (aq)	=O, O; + OH, C=CH–C–CH_3, OH	*138*
O, CH_3	=O, O; + OH, O, CH_3, O	

REFERENCES

1. A. Windaus and J. Brunken, *Ann. Chem.*, **460,** 225 (1928).

2. W. Fenical, D. R. Kearns, and P. Radlick, *J. Amer. Chem. Soc.*, **91,** 3396 (1969).

3. G. O. Schenck, H. D. Becker, K. H. Schulte-Elte, and C. H. Krauch, *Chem. Ber.*, **96,** 509 (1963).

4. H. Kautsky *Trans. Faraday Soc.*, **35,** 216 (1939).

5. D. B. Sharp, U.S. patent 2,951,797 (to Monsanto), 1960.

6. C. S. Foote and S. Wexler, *J. Amer. Chem. Soc.*, **86,** 3879 (1964).

7. C. S. Foote and S. Wexler, *J. Amer. Chem. Soc.*, **86,** 3880 (1964).

8. S. J. Arnold, N. Finlayson, and E. A. Ogryzlo, *J. Chem. Phys.*, **44,** 2529 (1966).

9. R. M. Badger, A. C. Wright, and R. F. Whitlock, *J. Chem. Phys.*, **43,** 4345 (1965).

10. S. J. Arnold, E. A. Ogryzlo, and H. Witzke, *J. Chem. Phys.*, **40,** 1769 (1964).

11. D. R. Kearns, R. A. Hollins, A. U. Khan, R. W. Chambers, and P. Radlick, *J. Amer. Chem. Soc.*, **89,** 5455 (1967).

12. D. R. Kearns, R. A. Hollins, A. U. Khan, and P. Radlick, *J. Amer. Chem. Soc.*, **89,** 5456 (1967).

13. K. Gollnick and G. O. Schenck, in *1,4-Cycloaddition Reactions* (J. Hamer, ed.), Vol. 8, Academic Press, New York, 1967, p. 255.

14. K. Gollnick, in *Advances in Photochemistry* (W. A. Noyes, Jr., G. S. Hammond, and J. N. Pitts, Jr., eds.), Vol. 6, Wiley-Interscience, New York, 1968, p. 1.
15. K. Gollnick and G. O. Schenck, *Pure Appl. Chem.*, **9,** 507 (1964).
16. C. S. Foote, *Account. Chem. Res.*, **1,** 104 (1968).
17. Y. A. Arbuzov, *Russ. Chem. Rev.* (English Transl.), **34,** 558 (1965).
18. R. Livingston, in *Autoxidation and Antioxidants* (W. O. Lundberg, ed.), Vol. 1, Wiley-Interscience, New York, 1961, p. 249.
19. E. J. Bowen, in *Advances in Photochemistry* (W. A. Noyes, Jr., G. S. Hammond, and J. N. Pitts, Jr., eds.), Vol. 1, Wiley-Interscience, New York, 1963, p. 23.
20. W. Bergman and M. J. McLean, *Chem. Rev.*, **28,** 367 (1941).
21. S. N. Foner and R. L. Hudson, *J. Chem. Phys.*, **25,** 601 (1956).
22. E. J. Corey and W. C. Taylor, *J. Amer. Chem. Soc.*, **86,** 3881 (1964).
23. R. W. Murray and M. L. Kaplan, *J. Amer. Chem. Soc.*, **90,** 537 (1968).
24. R. W. Murray and M. L. Kaplan, *J. Amer. Chem. Soc.*, **91,** 5358 (1969).
25. H. H. Wasserman and J. R. Scheffer, *J. Amer. Chem. Soc.*, **89,** 3073 (1967).
26. P. R. Story, D. D. Denson, C. E. Bishop, B. C. Clark, Jr., and J.-C. Farine, *J. Amer. Chem. Soc.*, **90,** 817 (1968).
27. E. J. Bowen and R. A. Lloyd, *Proc. Roy. Soc.* (*London*), *Ser. A*, **275,** 465 (1963).
28. J. A. Howard and K. U. Ingold, *J. Amer. Chem. Soc.*, **90,** 1056 (1968).
29. A. U. Khan and M. Kasha, *J. Chem. Phys.*, **39,** 2105 (1963).
30. R. J. Browne and E. A. Ogryzlo, *Proc. Chem. Soc.*, 117 (1964).
31. A. U. Khan and M. Kasha, *Nature*, **204,** 241 (1964).
32. E. McKeown and W. A. Waters, *J. Chem. Soc.* (*B*), 1040 (1966).
33. E. J. Bowen, *Pure Appl. Chem.*, **9,** 473 (1964).
34. E. J. Bowen and R. A. Lloyd, *Proc. Chem. Soc.* (*London*), 305 (1963).
35. G. O. Schenck, K. G. Kinkel, and H. J. Mertens, *Ann. Chem.*, **584,** 125 (1953).
36. R. N. Moore and R. J. Lawrence, *J. Amer. Chem. Soc.*, **80,** 1438 (1958).
37. A. Nickon and W. L. Mendelson, *J. Amer. Chem. Soc.*, **85,** 1894 (1963).
38. A. Nickon and W. L. Mendelson, *J. Amer. Chem. Soc.*, **87,** 3921 (1965).
39. C. Dufraisse and M. Badoche, *Compt. Rend.*, **200,** 929 (1935).
40. C. Dufraisse and M. Badoche, *Compt. Rend.*, **200,** 1103 (1935).
41. E. J. Bowen and D. W. Tanner, *Trans. Faraday Soc.*, **51,** 475 (1955).
42. E. J. Bowen and A. H. Williams, *Trans. Faraday Soc.*, **35,** 765 (1939).
43. E. J. Bowen, *Discussions Faraday Soc.*, **14,** 143 (1953).
44. R. Livingston and V. S. Rao, *J. Phys. Chem.*, **63,** 794 (1959).
45. E. J. Forbes and J. Griffiths, *Chem. Commun.*, 427 (1967).
46. K. R. Kopecky and H. J. Reich, *Can. J. Chem.*, **43,** 2265 (1965).
47. G. O. Schenck, *Ann. Chem.*, **584,** 156 (1953).
48. G. O. Schenck and D. E. Dunlap, *Angew. Chem.*, **68,** 248 (1956).
49. G. O. Schenck, K. Gollnick, and O. A. Neümuller, *Ann. Chem.*, **603,** 46 (1957).
50. G. O. Schenck, *Ind. Eng. Chem.*, **55,** 40 (1963).
51. G. O. Schenck and K. Ziegler, *Naturwiss.*, **32,** 157 (1944).
52. G. O. Schenck and R. Wirtz, *Naturwiss.*, **40,** 581 (1953).
53. G. O. Schenck, *Angew. Chem.*, **69,** 579 (1957).
54. W. D. Willmund, Disseration, University of Gottingen (1953).
55. W. R. Adams, and D. J. Trecker, unpublished results.
56. K. H. Schulte-Elte, B. Willhalm, and G. Ohloff, *Angew. Chem.*, *Int. Ed.*, **8,** 985 (1969).

57. G. O. Schenck, W. Müller, and H. Pfennig, *Naturwiss.*, **41,** 374 (1954).
58. C. Dufraisse, G. Rio, and J. J. Basselier, *Compt. Rend.*, **246,** 1640 (1958).
59. C. Dufraisse, A. Etienne, and J. J. Basselier, *Compt. Rend.*, **244,** 2209 (1957).
60. E. J. Forbes and J. Griffiths, *J. Chem. Soc.* (*C*), 601 (1967).
61. E. J. Forbes and J. Griffiths, *J. Chem. Soc.* (*C*), 572 (1968).
62. G. Rio and J. Berthelot, *Bull. Soc. Chim.*, 1664 (1969).
63. A. C. Cope, T. A. Liss, and G. W. Wood, *J. Amer. Chem. Soc.*, **79,** 6287 (1957).
64. G. Rio and M. Charafi, *Compt. Rend.*, **268,** 1960 (1969).
65. J. J. Basselier and J-P. LeRoux, *Compt. Rend.*, **268,** 970 (1969).
66. E. Demole and P. Enggist, *Helv. Chim. Acta*, **51,** 481 (1968).
67. S. H. Schroeter, R. Appel, R. Brammer, and G. O. Schenck, *Ann. Chem.*, **697,** 42 (1966).
68. E. Koch and G. O. Schenck, *Chem. Ber.*, **99,** 1984 (1966).
69. G. O. Schenck, *Angew. Chem.*, **56,** 101 (1944).
70. C. Dufraisse and S. Ecary, *Compt. Rend.*, **223,** 735 (1946).
71. H. H. Wasserman and A. R. Doumaux, Jr., *J. Amer. Chem Soc.*, **84,** 4611 (1962).
72. T. J. Katz, V. Balogh, and J. Schulman, *J. Amer. Chem. Soc.*, **90,** 734 (1968).
73. H. H. Wasserman and R. Kitzing, *Tetrahedron Lett.*, 5315 (1969).
74. C. S. Foote, M. T. Wuesthoff, S. Wexler, I. G. Burstain, R. Denny, G. O. Schenck, and K. H. Schulte-Elte, *Tetrahedron*, **23,** 2583 (1967).
75. J. Martel, *Compt. Rend.*, **244,** 626 (1957).
76. G. O. Schenck, *Chem. Ber.*, **80,** 289 (1947).
77. H. H. Wasserman and A. Liberles, *J. Amer. Chem. Soc.*, **82,** 2086 (1960).
78. G. O. Schenck, *Angew. Chem.*, **60,** 244 (1948).
79. G. O. Schenck, *Angew. Chem.*, **64,** 12 (1952).
80. G. O. Schenck and R. Appel, *Naturwiss.*, **33,** 122 (1946).
81. F. Bernheim and J. E. Morgan, *Nature*, **144,** 290 (1939).
82. P. de Mayo and S. T. Reid, *Chem. Ind.*, 1576 (1962).
83. W. Theilacker and W. Schmidt, *Ann. Chem.*, **605,** 43 (1957).
84. J. Auerbach and R. W. Franck, *Chem. Commun.*, 991 (1969).
85. G. Rio, A. Ranjon, O. Pouchot, and M. J. Scholl, *Bull. Soc. Chim.*, 1667 (1969).
86. H. H. Wasserman and A. H. Miller, *Chem. Commun.*, 199 (1969).
87. H. H. Wasserman and M. B. Floyd, *Tetrahedron*, Supplement No. 7, 441 (1966).
88. D. W. Kurtz and H. Shechter, *Chem. Commun.*, 689 (1966).
89. H. H. Wasserman and E. Druckrey, *J. Amer. Chem. Soc.*, **90,** 2440 (1968).
90. C. Dufraisse and J. Martel, *Compt. Rend.*, **244,** 3106 (1957).
91. J. Sonnenberg and D. M. White, *J. Amer. Chem. Soc.*, **86,** 5685 (1964).
92. E. H. White and M. J. C. Harding, *J. Amer. Chem. Soc.*, **86,** 5685 (1964).
93. H. H. Wasserman, K. Stiller, and M. B. Floyd, *Tetrahedron Lett.*, 3277 (1968).
94. T. Matsuura and I. Saito, *Tetrahedron*, **25,** 541 (1969).
95. T. Matsuura and I. Saito, *Tetrahedron*, **25,** 557 (1969).
96. T. Matsuura and I. Saito, *Chem. Commun.*, 693 (1967); T. Matsuura and I. Saito, *Tetrahedron*, **24,** 6609 (1968).
97. G. O. Schenck and C. H. Krauch, *Angew. Chem.*, 74, 510 (1962).
98. H. H. Wasserman and W. Strehlow, *Tetrahedron Lett.*, 791 (1970)
99. C. N. Skold and R. H. Schlessinger, *Tetrahedron Lett.*, 795 (1970).
100. C. Dufraisse and R. Priou, *Bull. Soc. Chim.*, **5,** 611 (1938).
101. C. Dufraisse and J. Houpillard, *Bull. Soc. Chim.*, **5,** 626 (1938).

102. C. Dufraisse and M. Gerard, *Compt. Rend.*, **201,** 428 (1935).
103. J. Rigaudy, C. Deletang, and J-J. Basselier, *Compt. Rend.*, **263,** 1435 (1966).
104. H. H. Wasserman and P. M. Keehn, *J. Amer. Chem. Soc.*, **88,** 4522 (1966).
105. A. Nickon and J. F. Bagli, *J. Amer. Chem. Soc.*, **81,** 6330 (1959).
106. A. Nickon and J. F. Bagli, *J. Amer. Chem. Soc.*, **83,** 1498 (1961).
107. G. O. Schenck, K. Gollnick, G. Buchwald, S. Schroeter, and G. Ohloff, *Ann. Chem.*, **674,** 93 (1964).
108. C. S. Foote, R. Wexler, and W. Ando, *Tetrahedron Lett.*, **46,** 4111 (1965).
109. K. Gollnick, S. Schroeter, G. Ohloff, G. Schade, and G. O. Schenck, *Ann. Chem.*, **687,** 14 (1965).
110. K. Gollnick and G. Schade, *Tetrahedron Lett.*, 2335 (1966).
111. W. R. Adams and D. J. Trecker, unpublished results.
112. C. S. Foote and J. W-P. Lin, *Tetrahedron Lett.*, 3267 (1968).
113. W. H. Schuller and R. V. Lawrence, *J. Amer. Chem. Soc.*, **83,** 2563 (1961).
114. C. S. Foote and M. Brenner, *Tetrahedron Lett.*, 6041 (1968).
115. W. F. Brill, unpublished results.
116. C. S. Foote and R. Denny, unpublished results.
117. R. Higgins, C. S. Foote, and H. Cheng, unpublished results.
118. G. Ohloff and G. Uhde, *Helv. Chim. Acta*, **48,** 10 (1965).
119. G. O. Schenck, H. Eggert, and W. Denk, *Ann. Chem.*, **584,** 177 (1953).
120. E. Klein and W. Rojahn, *Chem. Ber.*, **98,** 3045 (1965).
121. G. O. Schenck and K. H. Schulte-Elte, *Ann. Chem.*, **618,** 185 (1958).
122. J. A. Marshall and A. R. Hochstetler, *J. Org. Chem.*, **31,** 1020 (1966).
123. E. Klein and W. Rojahn, *Tetrahedron*, **21,** 2173 (1965).
124. H. G. Dässler, *Ann. Chem.*, **622,** 194 (1959).
125. R. A. Bell, R. E. Ireland, and L. N. Mander, *J. Org. Chem.*, **31,** 2536 (1966).
126. J. E. Fox, A. I. Scott, and D. W. Young, *Chem. Commun.*, 1105 (1967).
127. A. Nickon and W. L. Mendelson, *Can. J. Chem.*, **43,** 1419 (1965).
128. E. Klein and W. Rojahn, *Dragoco Rep.*, **14,** 95 (1967).
129. K. Gollnick and G. Schade, *Tetrahedron Lett.*, 689 (1968).
130. J. E. Huber, *Tetrahedron Lett.*, 3271 (1968).
131. N. Furutachi, Y. Nakadaira, and K. Nakanishi, *Chem. Commun.*, 1625 (1968).
132. M. Mousseron-Canet, J-C. Mani, and J. P. Dalle, *Bull. Soc. Chim.*, 608 (1967).
133. A. Nickon and W. L. Mendelson, *J. Org. Chem.*, **30,** 2087 (1965).
134. T. Matsuura and I. Saito, *Tetrahedron Lett.*, 3273 (1968).
135. E. Koerner Von-Gustorf, F. W. Grevels, and G. O. Schenck, *Ann. Chem.*, **719,** 1 (1969).
136. M. Mousseron-Canet, J. P. Dalle, and J-C. Mani, *Bull. Soc. Chim.*, 1561 (1968).
137. M. Mousseron-Canet, J. P. Dalle, and J-C. Mani, *Tetrahedron Lett.*, 6037 (1968).
138. S. Isoe, S. B. Hyeon, H. Ichikawa, S. Kutsumura, and T. Sakan, *Tetrahedron Lett.*, 5561 (1968).
139. S. Isoe, S. B. Hyeon, and T. Sakan, *Tetrahedron Lett.*, 279 (1969).
140. R. L. Kenny and G. S. Fisher, *J. Amer. Chem. Soc.*, **81,** 4288 (1959).

3

Epoxidation of Olefins by Hydroperoxides

RICHARD HIATT
DEPARTMENT OF CHEMISTRY
BROCK UNIVERSITY
ST. CATHARINES, ONTARIO, CANADA

I. Introduction . . . 113
A. Epoxidations by Hydrogen Peroxide . . . 113
B. Epoxidations by Hydroperoxides . . . 114
C. Metal Ion-Catalyzed Epoxidations . . . 116
II. Reaction Conditions . . . 117
A. Catalysts . . . 117
B. Hydroperoxides . . . 123
C. Olefins . . . 125
D. Concentrations and Solvents . . . 126
E. Temperature . . . 130
F. Effect of Oxygen . . . 130
G. Concurrent Autoxidation and Epoxidation . . . 131
III. Experimental Procedures . . . 131
IV. Kinetics and Mechanism . . . 134
References . . . 138

I. INTRODUCTION

A. Epoxidations by Hydrogen Peroxide

The synthesis of glycols by vanadium pentoxide- or chromium trioxide-catalyzed reaction of hydrogen peroxide with olefins was discovered by Milas and coworkers (*1*) in 1937 and subsequently confirmed by Triebs (*2*). Tungstic acid was also found to be an effective catalyst for the reaction

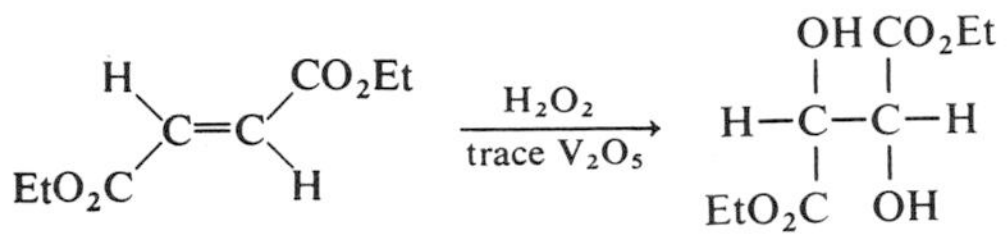

$$\text{(H)(EtO}_2\text{C)C=C(CO}_2\text{Et)(H)} \xrightarrow[\text{trace } V_2O_5]{H_2O_2} \text{H-C(OH)(CO}_2\text{Et)-C(OH)(CO}_2\text{Et)-H} \quad (57\%) \tag{1}$$

(*3–5*), which was proposed to proceed via the peracid of the metal (*1,2*).

$$V_2O_5 \xrightleftharpoons{H_2O_2 + H_2O} 2O{=}V(OH)_2{-}O{-}O{-}H \xrightarrow{\text{alkene}} \text{Products}$$

The blood-red color of pervanadic acid appeared when vanadium pentoxide dissolved in solutions containing hydrogen peroxide (*1*). Vanadium as well as a number of other metals (titanium, zirconium, thorium, niobium, tantalum, chromium, molybdenum, tungsten, and uranium) were known to form unstable peracids rather than peroxides when treated with hydrogen peroxide (*6*). Milas (*1a*) also showed that the combination would hydroxylate benzene, and obtained a 30% yield of phenol based on decomposed hydrogen peroxide.

More recently it has been found that the epoxide intermediates (leading to glycols) can be isolated in good yield. Vanadium pentoxide (*7*), tungsten trioxide (*8–11*), and molybdenum trioxide (*8c*) have been used, and yields as high as 97% have been claimed for the epoxidation of crotyl alcohol (*11*). Kinetic studies on a series of allyl alcohols are consistent with the proposed mechanism shown in Eq. (2).

$$WO_3 + H_2O_2 \longrightarrow O{=}W(OH)(O){-}O_2H \xrightarrow{CH_2{=}CH{-}CH_2OH} [\text{cyclic complex: } H_2C{=}CH{-}CH_2{-}O{-}W({=}O)_2{-}O{-}O(H)] \longrightarrow \text{Products} \tag{2}$$

B. Epoxidations by Hydroperoxides

Apart from the metal-catalyzed reaction (which is the only one of real synthetic potential), hydroperoxides may epoxidize olefins by at least three routes [Eq. (3)]. Of these, only the base-catalyzed addition to *alpha*,

$$RO_2H + \rangle C{=}C\langle \;\rightarrow\; \rangle \overset{O}{C{-}C}\langle + ROH \quad \text{or} \quad \rightarrow \text{Other products} \qquad (3)$$

beta-unsaturated aldehydes and ketones (*12,13*) offers efficient use of the hydroperoxide [Eq. (4)]. The method works, of course, only for compounds

$$RO_2^- + \rangle C{=}\overset{|}{C}{-}\overset{O}{\overset{\|}{C}}{-} \longrightarrow {-}\underset{RO{-}O^{\ominus}}{\overset{|}{C}}{-}\overset{|}{C}{-}\overset{O}{\overset{\|}{C}}{-} \longrightarrow {-}\overset{|}{C}{-}\overset{|}{C}{-}\overset{O}{\overset{\|}{C}}{-}\ (\text{epoxide, bridging } O) + RO^- \qquad (4)$$

undergoing Michael-type addition and is complicated by the possibility that protonation of the carbanion may be faster than its attack on O—O. Thus, *alpha*, *beta*-unsaturated nitriles and esters give dialkyl peroxides more frequently than epoxides (*12–14*).

Epoxides are produced in the autoxidation of simple olefins via addition of a peroxy radical to the double bond. Yields are usually low—1% for cyclopentene and cyclohexene (*15*) and 10–20% for simple acylic olefins (*16*)—due to other competing reactions (*15,16*).

$$RO_2\cdot + \rangle CH{-}\overset{|}{C}{=}C\langle \;\rightarrow\; \rangle CH{-}\dot{C}{-}C{-}OOR \;\longrightarrow\; \rangle CH{-}\overset{O}{C{-}C}{-} + RO\cdot$$

$$\rangle CH{-}\dot{C}{-}C{-}OOR \;\rightarrow\; \rangle CH{-}\overset{O{-}O\cdot}{\overset{|}{C}}{-}C{-}OOR \qquad (5)$$

$$RO_2\cdot + \rangle CH{-}\overset{|}{C}{=}C\langle \;\rightarrow\; RO_2H + \rangle\dot{C}{-}\overset{|}{C}{=}C\langle$$

Somewhat higher yields of epoxides are obtained by thermal decomposition of hydroperoxides in olefinic solvents (as high as 40–50% for *t*-butyl hydroperoxide and diisobutylene, based on decomposed hydroperoxide) (*17–19*). Although the reaction might be supposed to follow a peroxy radical mechanism as shown in Eq. (5), both Brill (*17*) and Walling and Heaton (*20*) have shown that not all the epoxide can arise via this route. A nonradical nucleophilic displacement by C=C on the oxygen–oxygen bond (**1**) analogous to epoxidation by peroxy acids (**2**) has been proposed (*20*).

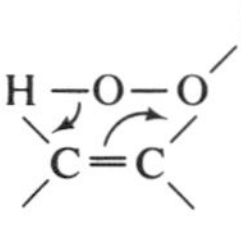

(**1**) 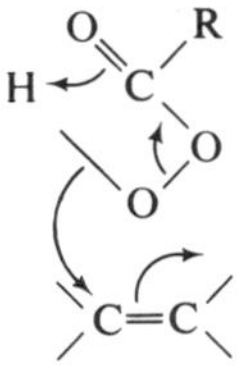 (**2**)

C. Metal Ion-Catalyzed Epoxidations

The earliest use of vanadium pentoxide to catalyze a hydroperoxide–olefin reaction appears to be that of Hawkins (*21*), who obtained "good" yields of epoxide from cyclohexene (39%) and 1-octene (about 15%). Much higher yields (53% from 1-octene and 100% from 2,4,4-trimethyl-1-pentene) were reported by Indictor and Brill (*22*), who used the oil-soluble acetylacetonates of either vanadium, chromium, or molybdenum. The discovery that propylene could be epoxidized by hydroperoxides generated *in situ* has led to a flurry of industrial interest and patents (*23–26*), the best known of which is probably the process developed at Halcon (*23*) [Eq. (6)].

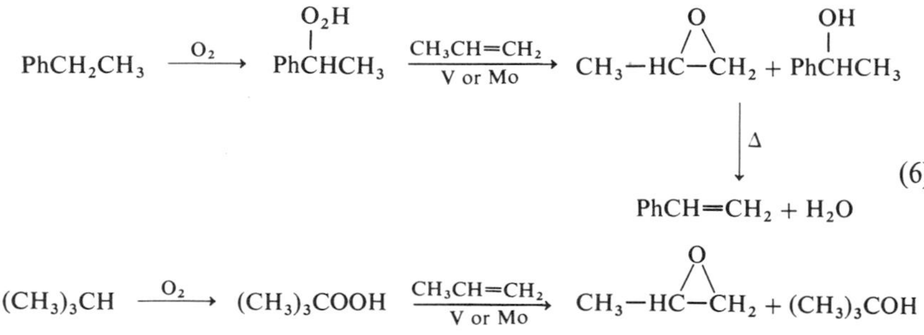

$$PhCH_2CH_3 \xrightarrow{O_2} PhCH(O_2H)CH_3 \xrightarrow[\text{V or Mo}]{CH_3CH{=}CH_2} CH_3{-}\overset{O}{HC{-}CH_2} + PhCH(OH)CH_3 \xrightarrow{\Delta} PhCH{=}CH_2 + H_2O \tag{6}$$

$$(CH_3)_3CH \xrightarrow{O_2} (CH_3)_3COOH \xrightarrow[\text{V or Mo}]{CH_3CH{=}CH_2} CH_3{-}\overset{O}{HC{-}CH_2} + (CH_3)_3COH$$

As yet there have been few publications in the nonpatent literature (*27–33*), and no reports of the reaction's use for routine synthesis of epoxides. In part, this may be due to the comment, probably overly pessimistic, of

Indictor and Brill (*22*) that "... the rates, yields and convenience on a preparative scale cannot compete with methods using peracids."

List and Kuhnen (*28*) have published a short but excellent review describing other reactions utilizing the vanadium or molybdenum–hydroperoxide combination. Thus, tertiary amines yield amine oxides (*34,35*) [Eq. (7)], and primary amines yield oximes (*23g,36*) [Eq. (8)]. Titanate esters also catalyze this reaction (*23g,36*).

$$R_3N + R'O_2H \xrightarrow{\text{Mo or V}} R_3N{\rightarrow}O + R'OH \quad (7)$$

$$RR'CHNH_2 + R''O_2H \xrightarrow{\text{Mo or V}} RR'C{=}NOH + ROH + H_2O \quad (8)$$

Sulfoxides are oxidized to sulfones. Attack on sulfur is faster than on the carbon–carbon double bond, as shown by the reaction of diallyl sulfoxide (*28,37*) [Eq. (9)].

$$(CH_2{=}CH{-}CH_2)_2SO \xrightarrow[\text{V or Mo}]{RO_2H} (CH_2{=}CH{-}CH_2)_2SO_2 \quad (9)$$

II. REACTION CONDITIONS

Briefly, **decompositions of hydroperoxides catalyzed by molybdenum or vanadium compounds in the presence of olefins can produce the corresponding epoxide in near-quantitative yields. Both soluble and insoluble catalysts can be used and the hydroperoxide may be generated *in situ*, the epoxidizing catalysis also serving to facilitate autoxidation. The olefin can, but need not be, used as the solvent; temperatures for preparative reactions are ordinarily 50–130°, with times ranging from several minutes to several days,** depending upon a variety of factors. Competing reactions such as decomposition of hydroperoxide by alternative routes or hydrolysis of the epoxide are negligible under optimum conditions.

The epoxide product is invariably one of *cis* addition to the carbon–carbon double bond (*22,27,29,30*). The hydroperoxide is in large part reduced to alcohol, although small quantities of products derived from RO· cleavage have also been observed (*23g,29*).

A. Catalysts

Catalytic activity for epoxidation by hydroperoxides has been claimed for compounds of a great many metals including chromium, molybdenum, niobium, nickel, rhenium, selenium, tantalum, titanium, vanadium, tungsten, and zirconium. Comparison studies (*22,25a,27,29,30*) (Table I)

TABLE I

Comparison of Catalysts for Epoxidation of Olefins by Hydroperoxides

Catalyst	Olefins	Hydroperoxide	Conditions	Epoxide yield, %[b]	Reference
None	2,4,4-Trimethyl-1-pentene	*t*-Butyl hydroperoxide	90°	45–50	*22*
Cr $acac_3$[a]			4–7 days at 25°[c]	100	
V $acac_3$				100	
MoO $acac_2$				100	
Co $acac_3$				30	
Cu $acac_2$				25	
Mn acac				15	
None	1-Octene, 2-octene mixture	Cumyl hydroperoxide	5–20 hr at 120°[d]	10	*29*
V_2O_5				7.3	
Nb_2O_5				7.8	
SeO_2				6.1	
CrO_3				2.2	
MoO_3				78.7	
WO_3				23.5	
None	2-Methyl-2-pentene	Cumyl hydroperoxide	15–0.5 hr at 110°[e]	29	*25a*
Na_2MoO_4				70	
Na_2WO_4				35	
Na_3VO_4				52	
MoO_2 $acac_2$			0.75–1 hr at 110°	86	
VO $acac_2$				75	

Co naph	$Me_2C(O_2H)CH{=}CH{-}CH_3$,	5–35 hr at 65° [f]	7.2	27
Cr naph	$Me_2C{=}CH{-}CH(O_2H){-}CH_3$ mixture		9.3	
MoO_3			42.9	
$NiCl_5$			27.1	
$ZrO_2 \cdot H_2O$			30.7	
H_3TaO_4			32.8	
Ti(O *n*-Bu)$_4$			37.7	
$W(CO)_6$			33.0	
VC			49.2	

[a] acac, Acetylacetonate; naph, naphthenate.

[b] Percentage yield based on decomposed hydroperoxide.

[c] 4×10^{-4} catalyst, about 2 *M* hydroperoxide in olefin solvent. Under these conditions the acetylacetonates of Al, Fe, Mg, Ni, TiO, Zn, and Zr gave less than 10% decomposition of the hydroperoxide and negligible epoxide.

[d] 0.1–1.0 g of catalyst in 30 g of hydroperoxide + 41 g of olefin.

[e] 0.005–0.02 g of catalyst in 2 ml of olefin + 2 ml of hydroperoxide + 1 ml of methanol. Under these conditions K_2PtCl_6, RoCl, $Cu(NH_3)_4SO_4$, $KAu(CN)_4$, and acetylacetonates of Mn, Fe, Co, and Ni gave less than 5% epoxide.

[f] Decomposition of an autoxidized mixture of 2′- and 4-methyl-2-pentene containing 20–23 wt % hydroperoxide and about 15% other oxygenated materials. Yields quoted are for epoxy alcohols as weight percent of total oxygenated compounds, after decomposition (see text).

have demonstrated that many, if not all, of these provide a higher yield of epoxide than uncatalyzed reactions. [Cobalt and manganese compounds generally produce less (*22,27,31,32*).] Only molybdenum, vanadium, and tungsten compounds appear to give consistently high epoxide yields, although much depends on optimization of conditions (Table I and below), and other metals could possibly be made equally effective.

Molybdenum compounds appear to be the best catalysts (*30*); those compounds most frequently used are molybdenum trioxide (*24,25,27,29, 30,32,33b*), sodium molybdate (*25,30*), molybdic acetylacetonate (*22,24,25, 27,30,32*), molybdenum naphthenate (*23,25*), and molybdenum hexacarbonyl (*23f,25,30*). Sheng and Zajacek (*25,30*) have published the most extensive survey including (as well as the above) molybdenum disulfide, molybdenum dioxide, molybdenum pentachloride, silicomolybdic acid, phosphomolybdic acid, sodium silico-12-molybdate, sodium phospho-12-molybdate, sodium phospho-18-molybdate, ethyl phosphomolybdate, monoethyldihydrophosphomolybdate, and molybdenum metal. Under standard conditions (Table I, Ref. *25a*), most of these, alone or in combinations, gave 80–88% yields of epoxide, the exceptions being molybdenum disulfide (52%), silicomolybdic acid (53%), phosphomolybdic acid (59%), sodium phosphomolybdate (63%), and sodium molybdate (70%) alone.

Under slightly different conditions (0.1 g of catalyst, 100° for 30 hr), 92% yields could be achieved with sodium molybdate. However, the best catalysts of this type appeared to be mixtures of sodium molybdate with sodium silico-12-molybdate with 0.05 g of catalyst providing a 96% conversion of cumene hydroperoxide in 2 hr at 100°, giving a 98% yield of 2-methylpentene-2-epoxide. In their rate studies using molybdenum hexacarbonyl and lower conversions and concentrations of reactants, Sheng and Zajacek (*30*) have reported yields of epoxide to be quantitative.

From the above it might appear that, given optimum conditions, any molybdenum compound could produce nearly quantitative yields of epoxide. Sheng and Zajacek (*25a*), however, express a preference for the hetero-12-molybdates such as sodium phospho-12-molybdate and sodium silico-12-molybdate over the 2-hetero-18-molybdates (sodium phospho-18-molybdate, etc.), which they say tend to destroy hydroperoxides by other routes.

Apparently, compounds that are only slightly soluble or insoluble in organic solvents, such as molybdenum metal, molybdenum trioxide, or the inorganic molybdates, can be just as effective catalysts as the oil-soluble acetylacetonates, naphthenates or carbonyls. It is uncertain, however, whether such catalysis is heterogeneous or results from small amounts of dissolved material, perhaps in some cases formed by reaction with the hydroperoxide. The latter is implied by Sheng and Zajacek (*25,30*), and certainly has an analogy in the formation of soluble pervanadates from hydrogen peroxide and vanadium pentoxide (*1*) and permolybdates from molybdenum trioxide and hydrogen peroxide (*29*).

Mashio and Kato (*29*) found no simple relationship between the rate of reaction and the amount of solid catalyst added; nor does there appear to be any special need for surface activation (*30*). Gould and Rado (*32*), however, argue that the amount of material in solution under their experimental conditions could not possibly give the observed reaction rates. Thus, it must be heterogeneous.

Examples of vanadium compounds used as catalysts include vanadium pentoxide (*23e,28,29,32*), sodium vanadate (*25a,30*), vanadium acetylacetonate (*22,31*), vanadyl acetylacetonate (*22,24,25a,27,30–32*), vanadium naphthenate (*23,24,27*), sodium metavanadate (*27*), vanadium carbide (*27*), and vanadium metal (*27*). High to virtually quantitative yields of epoxide have been reported for the more reactive olefins such as cyclohexene (*23e, 31,32*), methylpentenes (*22,25a,27*), and allyl alcohols (*28*). However, in the epoxidation of propylene, vanadium naphthenate is notably inferior to the molybdenum compound, giving respectively 38% (*23e*) and 86% (*23f*) yields (based on conversion of hydroperoxide) under similar conditions, whereas vanadium pentoxide produced only 6%. Mashio and Kato (*29*) also obtained lower yields with vanadium pentoxide than with no catalyst at all in the reaction of *t*-butyl or cumene hydroperoxide with 1-octene (Table I), but this was probably due to the high temperature (120°) employed.

Good yields have been reported from use of a vanadium catalyst only for temperatures below 90°, and Allison et al. (*27*) have shown that epoxide yields (with vanadium naphthenate) fall off sharply above this temperature. It is not known whether this results from destruction of the epoxide by further reaction or by side reactions between catalyst and hydroperoxide. Kollar et al. (*23e*) have reported increased yields from vanadium-catalyzed reactions when a small amount of sodium naphthenate or other mild base was added. However, acids could conceivably catalyze decomposition of either hydroperoxide or epoxide.

Comparison of kinetic studies on molybdenum (*30*) and vanadium (*31*) catalysis suggests that the latter is much more subject to autoinhibition, both from the product alcohols and from oxidative degradation of the catalyst.

Tungsten compounds, though apparently the best catalysts in combination with hydrogen peroxide (*3–11*), rank far below both molybdenum and vanadium in general use and usually in yields, where reported (*25a,29,30*) (Table I). However, Allison and coworkers (*24*) have found tungsten hexacarbonyl to give epoxide yields from the decomposition of partly oxidized 2- and 4-methyl-2-pentenes comparable to those obtained from molybdenum hexacarbonyl. Rouchaud et al. (*33*) obtained 50% higher conversion of propylene oxide using tungsten trioxide rather than molybdenum trioxide, although from their reported conditions it is not clear whether the oxidant was hydroperoxide or organic peracid.

Although Indictor and Brill's (*22*) original publication dealt mainly with the high epoxide yields obtainable from chromium triacetylacetonate in a very slow reaction at 25° (Table I), they reported that the yields dropped drastically with increased temperature. Subsequent reports by other workers (*27,29,32*) have emphatically confirmed this temperature dependence but have shed no light on the reasons. It has been claimed (*23g*) that chromium trioxide in combination with a hydroperoxide is excellent for converting *alpha,beta*-unsaturated aldehydes to the corresponding carboxylic acids.

The only support for the usefulness of other metals as catalysts, apart from patent claims, is the work of Allison et al. (*27*). Unfortunately this study defies clear interpretation since the decompositions were carried out on an incompletely analyzed, partially oxidized mixture of 2- and 4-methyl-2-pentene (Table I) with no uncatalyzed reaction as a control. The products included epoxy alcohols and 20–25% of "oxide," even from the cobalt naphthenate-catalyzed decomposition. This "oxide" is presumably olefin epoxide, although from the ambiguity of the text it might equally well be mesityl oxide. The epoxy alcohol yields, however, from niobium pentachloride, zirconium dioxide, tantalic acid, and tetra-*n*-butyl orthotitanate-catalyzed reactions equaled those obtained from the use of molybdenum or vanadium compounds.

The alkyl orthotitanates appear to be excellent catalysts for the related hydroperoxide-induced oxidations of sulfoxides to sulfones (*28,37*) and primary amines to oximes (*23g,36*). **Dimethyl sulfoxide, for example, yields 94% of dimethyl sulfone when treated with cumene hydroperoxide containing 0.5 wt% of tetraisobutyl titanate** (*28*).

Boric acid has been reported to promote the formation of *trans*-diols from olefins with cumene hydroperoxide (*38*), a 67% yield being claimed for the preparation of 1,2-dodecanediol from dodecene. The probability of an epoxide intermediate is demonstrated by a recent publication showing that equimolar quantities of alkyl metaborates (*39*) and hydroperoxide produce epoxides quantitatively from olefins. Though this reaction is not "catalytic," the difference is only that metaborate is apparently transformed in the reaction to the orthoborate, which is inactive. The metaborate–hydroperoxide combination will also hydroxylate mesitylene in good yield, but not benzene or toluene.

B. Hydroperoxides

From the available literature, it appears that the choice of hydroperoxide is the least critical factor. Table II shows examples of typical conditions and yields. In the two cases where yields are conspicuously low, there seems to have been no attempt to optimize conditions (*25a*). In all of the examples, the hydroperoxides were added or preformed in known concentration prior to catalytic decomposition. Instances of concurrent autoxidation and epoxidation will be discussed in a later section. Ketone hydroperoxides such as 2-hydroxy-2-butyl hydroperoxide are claimed to be as effective as *t*-butyl or cumyl hydroperoxide (*23e*), although yields are not reported.

Clearly the purity of the hydroperoxide is not a critical factor either (Table II). This is particularly interesting in the case of the commercial-grade 72% *t*-butyl hydroperoxide (Table II, Ref. *25a*) which, since it probably contained considerable water, might be expected to produce more diol than epoxide. Nonaqueous hydrogen peroxide, as shown, yields more diol than epoxide under conditions in which alkyl hydroperoxides produce epoxide exclusively (*29*).

Table III shows the effect of hydroperoxide structure on the rate of catalyzed epoxidation. The rate-enhancing effect of an electron-withdrawing group (which presumably reduces electron density on the peroxidic oxygens) (*30*) is apparent. Steric effects may also be important. For example, Allison et al. (*27*) have reported that 4-methyl-2-buten-4-yl hydroperoxide and 4-methyl-3-buten-2-yl hydroperoxide react at considerably different rates.

TABLE II

Comparison of Hydroperoxides

Hydroperoxide	Olefin	Catalyst (wt % of hydroperoxide)	Conditions	Unreacted hydroperoxide, %	Yield of epoxide, %[a]	Reference
Hydrogen peroxide	Cyclohexene	(10) MoO_3	10 hr at 50° in dioxane	7.6	37.6[j]	*29*
t-Butyl hydroperoxide		(12) MoO_3	8 hr at 80° in dioxane	10	99	*29*
2-Cyclohexenyl hydroperoxide		(2) MoO_2 $acac_2$[b]	1 hr at 70°	1.4	86	*32*
Cumyl hydroperoxide		(0.03) V naph	1 hr at 90° in ethylbenzene	3	97	*23e*
5-Butyl hydroperoxide	Cyclopentene	(1) V octoate	6 hr at 46°	15	98	*31*
t-Butyl hydroperoxide[c]	1-Octene	(8) Molybdates[d]	2 hr at 135° in ethylacetate	0	89	*24a*
Cumyl hydroperoxide		(19) MoO_3	6 hr at 100° in cumene	10	84	*29*
t-Amyl hydroperoxide[e]	Propene	(2) Mo naph	2 hr at 130°		78	*23g*
Cumyl hydroperoxide[f]	2-Me-2-pentene	(1.5) Molybdates[d]	0.75 hr at 110°	10	97	*25a*
2,4-Dimethyl-3-keto-2-pentyl hydroperoxide[h]		(3) Molybdates[g]	1 hr at 85° in methanol	45	43	*25a*
		(1.5) Molybdates[g]	1 hr at 75° in methanol	16	100	*25a*
1-Tetralin hydroperoxide[i]		(3.5) Molybdates[g]	1 hr at 85° in chloroform	16	20	*25a*

[a] acac, Acetylacetonate; naph, naphthenate. All peroxides better than 97% pure unless otherwise indicated.
[b] Based of reacted hydroperoxide.
[c] 72% pure.
[d] Equal weight of Na_2MoO_4 and $Na_4(PMo_{12}O_{40})$.
[e] 50-50-hydroperoxide–alcohol.
[f] 82% pure.
[g] Equal weights of Na_2MoO_4 and $Na_4(SiMo_{12}O_{40})$.
[h] 89% pure.
[i] 74% pure.
[j] Together with 54.5% yield of *trans*-diol.

TABLE III

Rate Dependence on Hydroperoxide Structure

	Relative rates in several olefins		
Hydroperoxide	2-Octene, 80.5°[a]	2-Methyl-1-pentene, 80.5°[a]	Cyclohexene, 80.0°[d]
t-Butyl hydroperoxide	1.00	1.00	1.00[e]–1.00[f]
Cumyl hydroperoxide	1.18[b]–1.21[c]	1.32[b]–1.15[c]	1.11[e]–1.07[f]
p-Nitrocumyl hydroperoxide	2.28[b]–1.67[c]		

[a, b, c] Calculated from conversions of hydroperoxide after [b]15 min and [c]30 min (Ref. *30*), assuming a pseudo-first-order reaction. Conversions ranged from 48 to 92%; catalyst, molybdenum hexacarbonyl; epoxide yields stated to be quantitative.

[d, e, f] From second-order rate constants (rate $= k_2[RO_2H][Olefin]$) in [e]dioxane or [f]cumene solvent; catalyst molybdenum trioxide (Ref. *29*).

C. Olefins

Olefins for which, on a preparative scale, numerical yields of epoxide have been reported (generally 70–90% based on decomposed hydroperoxide) include propene (*23,25,33*), *cis-* and *trans*-2-butene (*25a,30*), isobutene (*25a*), 2-methyl-2-butene (*25a*), 2-ethyl-1-butene (*27*), 2-methyl-1-pentene (*25a*), 2-methyl-2-pentene (*25a,27*), 3-methyl-2-pentene (*27*), *cis-* and *trans*-4-methyl-2-pentene (*27,29*), 1-hexene (*23f*), 2-hexene (*27*), 1-octene (*22,29,30*), 2-octene (*22,25a,29,30*), 2,4,4-trimethyl-1-pentene (*22*), 1-dodecene (*29*), 1-tetradecene (*29*), cyclopentene (*31*), cyclohexene (*23,25a,26,27,29,30–32*), 4-vinylcyclohexene (yielding 4-vinyl-1,2-epoxycyclohexane) (*22,25a*), 1,3-cyclooctadiene (*25a,26*), 1,5-cyclooctadiene (*25a*), and a low-molecular-weight stryene–butadiene copolymer (*29*).

Patent claims are, of course, extensive and cover generically virtually all olefinic substances, including high-molecular-weight polymers, both conjugated and unconjugated dienes, and compounds containing other functional groups such as hydroxyl, halogen, cyano, carboxyl, ester, ether, and keto (*23e,25c*). It appears, however, that the only example of nonhydrocarbons for which yields have been published is a series of allylic alcohols (*28*) which, with cumene hydroperoxide and vanadium pentoxide, at 60° gave 60–95% of the corresponding epoxy–alcohol.

Notwithstanding the above, Mashio and Kato (*29*) obtained little or no epoxide from styrene or *alpha*-methylstyrene under conditions (*t*-butyl hydroperoxide plus molybdenum trioxide at 100°) in which simple olefins gave good yields. Instead, the products were an approximately 4 : 1 mixture of *alpha*-phenylpropionaldehyde and acetophenone, both apparently resulting from *in situ* decomposition of the epoxide.

The effect of structure on olefin reactivity appears to parallel that for epoxidation by organic peracids (*40*). That is, increasing alkyl substituents bonded to the carbons of the double bond result in enhanced rates of reaction (*30*). However, the confirmatory evidence (Table IV) is as yet mainly qualitative and rather sparse. Clearly, relative reactivity varies somewhat with the catalyst and the hydroperoxide (Table IV). *cis*- and *trans*-2-Octenes have been shown to react with essentially equal rates [Table IV, *cis-trans* isomerization was shown to be absent (*29*)], but no other comparisons of *cis* and *trans* isomers are available. Measurements by Sheng and Zajacek (*35*) show a relative reactivity for 1-octene/2-octene of 2.56 (*t*-butyl hydroperoxide/molybdenum hexacarbonyl at 85°).

Based on this limited evidence, **it appears that metal ion/hydroperoxide reagents are less sensitive than organic peracids toward olefin structure, with the metaborate–hydroperoxide combination falling between the two in selectivity** (Table V). In part, of course, this may be due to the much lower temperatures ordinarily used for peracid epoxidation, but there is some evidence that metal ion/hydroperoxide reagents are not very discriminating even at 25° (Table IV).

D. Concentrations and Solvents

In most instances, an excess of olefin to hydroperoxide of at least 2 : 1 or 3 : 1, with the olefin frequently serving as solvent, is recommended. Lowering the olefin concentration, other things being equal, results in less efficient use of hydroperoxide, and the effect tends to be more pronounced the higher the catalyst concentration (*29,31*). Published reports to date, however, have concentrated solely on maximizing the yield based on hydroperoxide consumption. None, it appears, has been concerned with obtaining high conversions of olefin. In this connection it is promising to note that with rare exceptions (*29*) the only product obtained from the olefin is the epoxide.

Solvents, if used, generally constitute no more than one-half of the reaction volume. Sheng and Zajacek (*25a,30*) have reported the most extensive survey of solvents. Their compilations are summarized in Table VI. The difference in the first set of figures might be due, in part, to

TABLE IV

Relative Reactivities of Olefins Toward Epoxidation by Hydroperoxides and Metal Ions

Olefin	Conversion of Hydroperoxide[a]				Epoxide yield[b]
	t-Butyl hydroperoxide[c] Cr $acac_3$[d]	*t*-Butyl hydroperoxide[e] Mo $(CO)_6$	*t*-Butyl hydroperoxide[f] Mo $(CO)_6$	Cumyl hydroperoxide[g] Na_2MoO_4	Cumyl t hydroperoxide[h] MoO_3
1-Octene	53	36			26
2-Octene	78	76	54	23	35
2-Methyl-2-pentene				52	
2-Methyl-1-pentene			39		
2,4,4-Trimethyl-1-pentene	100				
Cyclohexene		77		30	

[a] From parallel experiments with partial decomposition of hydroperoxide in excess of olefin under conditions where $-\Delta RO_2H = \Delta$ Epoxide.

[b] From complete decomposition of 0.184 mole of hydroperoxide in 0.368 mole of mixed octenes, 49% 1-octene, 26% *cis*-2-octene, 25% *trans*-2-octene.

[c] 4 days at 25° (Ref. *22*).

[d] acac, acetylacetonate.

[e] 30 min at 101° (Ref. *30*).

[f] 15 min at 80.5°, benzene solvent (Ref. *30*).

[g] 5 hr at 101° (Ref. *30*).

[h] 5 hr at 120° (Ref. *29*).

differences in solubility of the molybdate catalyst, but similar and rather more dramatic differences are obtained with the soluble catalyst, molybdenum hexacarbonyl.

It appears that solvents may affect the rate of reaction considerably without greatly influencing the efficiency of conversion (Table VI, columns 1 and 2). This is evident also for molybdenum hexacarbonyl–*t*-butyl hydroperoxide epoxidations of propylene in *t*-butyl alcohol–benzene mixtures (*25b,30*), although yields do fall off at *t*-butyl alcohol–benzene ratios greater than four. Mashio and Kato (*29*) have found the molybdenum trioxide–hydroperoxide–cyclohexane reaction to be about six times as fast in cumene as in dioxane, although they obtained quantitative yields of epoxide in either solvent.

TABLE V

Comparison of Hydroperoxide–Molybdenum, Hydroperoxide–Metaborate, and Peracid Epoxidations

Reagent	Temperature	Relative rate		Reference
		1-Octene	2-Octene	
Peracetic acid in acetic acid	25.8°	1	20–24[a]	*40*[a]
Peracetic acid in ethyl acetate	25.8°	1	35.6	*39*
1-Tetralin hydroperoxide/ cyclohexyl metaborate	80.0°	1	9.5	*39*
t-Butyl hydroperoxide/ molybdenum hexacarbonyl	85.5°	1	2.56	*30*

[a] Generalization from reactivities of corresponding hexenes, heptenes, etc.

Aliphatic and aromatic hydrocarbons clearly are better solvents for the reaction than alcohols (*25,30*); yet a certain amount of alcohol must be accommodated since it is produced in the course of the reaction and may be present initially, as well, if unpurified hydroperoxide is used (*23g,27*). This seems to be an inadequate reason for adding alcohol purposely, however, as has been done in a number of instances (*23e,23f,25,26b*) without explanation.

The rate-retarding effect of alcohols, which is much more severe for vanadium (*30*) than for molybdenum (*30*), is probably due to competition with hydroperoxide for complexing sites on the metal ion (*31*). Acetone

TABLE VI

Effect of Solvents on the Molybdenum-Catalyzed Epoxidation of 1-Octene by *t*-Butyl Hydroperoxide (25*a*,30)

Solvent	Unreacted hydroperoxide, %[a]	Yield of epoxide, based on reacted hydroperoxide, %[a]	Unreacted hydroperoxide, %[b]	Yield of epoxide, based on reacted hydroperoxide, %[b]
1-Octene	—	88	—	—
Ethyl acetate[d]	0	89.3	—	—
Diisopropyl ketone	1.3	89.2	—	—
Toluene	13.6	88.1	—	—
Acetophenone	1.4	68.8	—	—
Diethyl carbonate	24.5	88.2	—	—
Benzene	1.0	63.6	75	92
t-Butyl alcohol	1.2	61.9	16	58
Ethyl orthoformate	32.7	63.7	—	—
Methanol	30.4	60.9	—	—
Acetone	49.6	47.2	39	5
Dioxane	70.4	71.7	—	—
Methylcyclohexane	—	—	72	84
Ethanol	—	—	40	37

[a] 4-ml samples containing 0.9 ml of 72% hydroperoxide, 1.8 ml of olefin, 1.3 ml of solvent, and 0.02 g each of Na_2MoO_4 and $Na_3(PMo_{12}O_{40})$ heated for 2 hr at 135°.

[b] 1 g of 97% hydroperoxide, 3 ml of olefin, 3 ml of solvent, and 0.005 g of $Mo(CO)_6$ for 1 hr of 90°.

[c] Assuming no unreacted hydroperoxide.

[d] Contained half the amount of catalyst.

may act similarly. It is otherwise difficult to explain why it is so inferior to diisopropyl ketone or acetophenone as a reaction solvent (*25a,30*) (Table VI).

The concentrations of catalyst employed have varied from 0.1–1% by weight of hydroperoxide for soluble materials to 1–10% for insoluble or slightly soluble substances (Table II is illustrative).

There are no published studies of yields *vs.* catalyst concentration (as the sole variable), but Sheng and Zajacek (*25*) assert that increasing concentrations of metal ion promote side reactions.

E. Temperature

In general, it appears that the lower the temperature, the less hydroperoxide decomposes by other routes. Again, Sheng and Zajacek (*30*) have provided a systematic study for molybdenum catalysis showing that yields (based on hydroperoxide conversion) which may be as high as 98% at 100° fall to 75–80% at 130°. For vanadium a similar falling off has been noted above 80° (*27,29*), and as mentioned earlier, with chromium, the falling-off point must be below 60° (*22*). Tungsten, however, may require a higher temperature than any of these, Rouchaud et al. (*33*) having obtained maximum yields from tungsten trioxide-catalyzed epoxidation of propylene at 150°.

As evident from the examples in Tables II and VI, most syntheses using vanadium catalysis have involved temperatures of 60–90°, and those using molybdenum, 90–120°. However, the wide diversity of conditions and compounds used makes it impossible to judge which of these two types of catalysts would react faster at a given temperature (see Section IV).

F. Effect of Oxygen

Indictor and Brill found oxygen to completely inhibit chromium acetylacetonate-catalyzed epoxidations (*22*). But catalysts of vanadium and molybdenum appear to function perfectly well in an oxygen environment (see below), although some specific compounds [such as vanadium acetylacetonate (*31*)] undergo rather rapid oxidative degradation. Sheng and Zajacek (*30*) found the epoxidation of 1-octene by *t*-butylhydroperoxide–molybdenum hexacarbonyl to be completely unaffected by oxygen, although the catalyst, recovered and reused four times successively, gradually deteriorated and lost activity. This did not occur in an oxygen-free atmosphere.

G. Concurrent Autoxidation and Epoxidation

Several studies have been reported in which an olefin [cyclohexene (*26a,27,29,32*), 2- and 4-methyl-2-pentenes (*27*), propylene (*33*)] was treated with oxygen in the presence of molybdenum (*26a,27,29,32,33*), vanadium (*27,32*), tungsten (*27,33*), and other (*27,32*) metal catalysts [Eq. (13)].

$$\mathrm{{>}C{=}C{<}\!-\!CH{<}} \xrightarrow[\text{cat.}]{O_2} \mathrm{{>}C{=}C{<}\!-\!CO_2H{<}} \xrightarrow{\mathrm{{>}C=C-CH{<}}} \mathrm{{>}\overset{O}{\widehat{C-C}}{<}\!-\!CH{<}} + \mathrm{{>}C{=}C{<}\!-\!COH{<}} \qquad (13)$$

The several findings for cyclohexene are compared in Table VII. Apparently, at low oxygen pressure, efficient epoxidation occurs, even at rather high conversion. High oxygen pressure yields ketones at the expense of the epoxide. The reasons for this are not fully obvious; nor is it clear why Allison and coworkers obtained such a large measure of epoxy alcohol from cyclohexene and from other olefins (*27*) [Eq. (13a)]. The careful work of Machio and Kato (*29*) makes it clear that these are tertiary products formed only at high conversions, not from intramolecular rearrangement of the allylic hydroperoxide.

$$\text{cyclohexene} \longrightarrow \longrightarrow \text{2-cyclohexen-1-ol (OH)} \xrightarrow[\text{cat.}]{\text{cyclohexenyl-}O_2H} \text{2,3-epoxycyclohexanol (OH, O)} + \text{2-cyclohexen-1-ol (OH)} \qquad (13a)$$

A survey of other metal catalysts (oxides and acetylacetonates of copper, chromium, cobalt, manganese, and iron) by Gould and Rado (*32*), under the conditions shown in Table VII, showed that none gave more than 1–2 % epoxide. Chromium and iron produced high yields of cyclohexenone, while cobalt, copper, and manganese tended to yield about equal amounts of cyclohexanone and cyclohexanol.

III. EXPERIMENTAL PROCEDURES

General Procedure for Liquid Olefins *(25a)*

In a 1-liter autoclave were placed 100 ml of cumene hydroperoxide (technical grade, 82 % pure), 140 ml of olefin, 60 ml of methanol, 1.2 g of

sodium molybdate, and 0.3 g of sodium silico-12-molybdate. The mixture was stirred and heated at 110° for 2 hr, cooled, filtered, and fractionally distilled. Epoxide yields from olefins (based on hydroperoxide) were isobutene, 56.0%; 2-methyl-2-butene, 80.5%; 2-methyl-1-pentene, 63.2%; 2-octene, 72.7%; cyclohexene, 76%; 4-vinyl cyclohexene, 68.4% (1,2-epoxy-4-vinylcyclohexane).

Epoxidation of Propylene *(23g)*

A 100-g sample of a 50-50 *t*-amyl hydroperoxide–*t*-amyl alcohol mixture (obtained from autoxidation of isopentane) plus 168 g of propylene and 1.0 g of a molybdenum naphthenate solution, containing 5% molybdenum by weight, were heated in a pressure vessel at 130° for 2 hr. Fractional distillation of the products at atmospheric pressure (using a five-plate column) gave 150 g of recovered propylene, 22 g of propylene oxide, 4.0 g of methyl isopropyl ketone, and 84 g of *t*-amyl alcohol. The propylene oxide yield was 78%, based on reacted hydroperoxide.

In another example (*30*), 100 ml of propylene, 25 g of 94% *t*-butyl hydroperoxide, 125 g of benzene, and 0.05 g of molybdenum hexacarbonyl were heated and stirred in an autoclave at 105° for 1 hr. Analysis of the products by iodometric titration (for hydroperoxide) and gas chromatography showed 92% conversion of hydroperoxide and 86% yield of epoxide, based on hydroperoxide conversion.

Epoxidation of 1-Octene *(30)*

One gram of 97% *t*-butyl hydroperoxide, 3 ml of 1-octene, and 0.005 g of molybdenum hexacarbonyl were sealed in a pressure tube and heated at 90° for 1 hr. Analysis by iodometric titration and gas chromatography showed 75% conversion of hydroperoxide and a 92% yield of 1,2-epoxyoctane.

Epoxidation of Methallyl Alcohol *(28)*

A 1 : 2 : 1 mixture of cumene hydroperoxide, methallyl alcohol, and vanadium pentoxide was heated at 60° for 7 hr. Analysis showed 98.8% of the hydroperoxide consumed and an 86% yield of epoxide.

Comments on Published Procedures. Publications to date make it quite clear that metal-catalyzed epoxidations by hydroperoxide offer an inexpensive and practical means for producing large quantities of material.

TABLE VII

Autoxidation of Cyclohexene in the Presence of Epoxide-Forming Catalysts

Catalyst	Molarity or wt % olefin	Temp.	Time, hr	Conversion, mmole O_2/g olefin	Products[a]					Reference
					 cyclohexene oxide	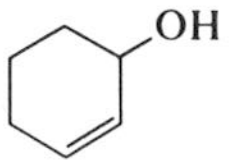 2-cyclohexen-1-ol	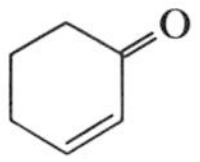 2-cyclohexen-1-one	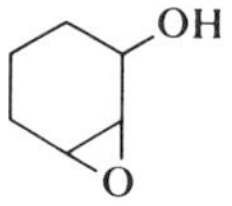 2,3-epoxycyclohexan-1-ol	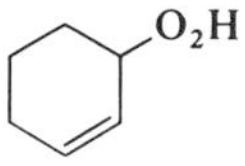 2-cyclohexenyl hydroperoxide (O_2H)	
VO $acac_2$[c]	0.001 *M*	70[d]	3.4	0.71	35.5	34.7	21.0		2.8	*32*
V naph	0.006 *M*	55[e]	15	—[b]	62			48	7.2	*27*
V_2O_5	6.2	70[d]	0.68	0.51	29.7	28.3	22.2		4.0	*32*
MoO_2 $acac_2$	0.001 *M*	70[d]	1.5	0.40	64.2	58.0	60.9		2.65	*32*
MoO_3	6.2	70[d]	1.7	0.19	60.1	63.0	27.1		2.6	*32*
MoO_3	0.1	70[f]	—	1.45	87	72	7			*26a*
MoO_3	2.4[g]	60[g]	3	0.274	90.1	84.6			2.7	*29*
MoO_3	1.2[g]	60[g]	12	1.0	73.9	63.2	13.5	6.8	2.2	*29*
ABN	0.001 *M*	70[d]	0.85	0.69	2.0	0	5.5		76.3	*27*

[a] Mmoles/100 mmoles O_2 uptake.
[b] 15% olefin conversion; would equal 0.9 mmole oxygen/g if each oxygen converted two molecules of cyclohexenes.
[c] acac, Acetylacetonate; naph, naphthenate.
[d,e,f] Oxygen pressure: [d] 55–95 psi; [e] 1 atm; [f] 100 mm.
[g] Also contained 0.02 *M* ABN (azo-bis-isobutyronitrile).

For a convenient laboratory procedure, however, competitive with epoxidations by organic peracids, the method is at this time promising but not fully proved. For a truly competitive procedure one might specify the following: (1) use of commercially available hydroperoxides and catalysts; (2) high yields based on olefin, rather than hydroperoxide, conversion; and (3) nonpressure reaction conditions—not sealed tubes or autoclaves.

A recent laboratory experiment employing nonpressure conditions provided the following results. Refluxing cyclohexene with an equimolar quantity of 90% *t*-butyl hydroperoxide and a bit of molybdenum naphthenate or molybdenum trioxide readily converts about half of the cyclohexene to cyclohexene oxide in 2–3 hr, while reducing 90–95% of the hydroperoxide to *t*-butyl alcohol. Some cyclohexanediol is also produced. It is possible to react all of the cyclohexene by starting with a higher ratio of *t*-butyl hydroperoxide, but under these conditions the cyclohexene oxide yield goes through a maximum (again about 50%) and then declines to nearly zero by the time all of the cyclohexene is consumed. Similar experiments using either benzene or pentane as solvent have given similar results. With pentane the optimum reaction time was increased to about 24 hr because of the lower reflux temperature.

It appears that considerably more development work, will be necessary to make the reaction competitive with the peracid method for routine laboratory synthesis.

IV. KINETICS AND MECHANISM

The reaction of hydroperoxide with molybdenum or vanadium compounds in the presence of olefins bears little resemblance to the better known type of hydroperoxide decomposition catalyzed by such metals as cobalt, manganese, and iron. These latter metal ions generate free radicals by a one-electron redox reaction with the hydroperoxides giving as typical products, oxygen, aldehydes, and ketones, as well as

$$RO_2H + Co^{2+} \longrightarrow RO\cdot + [Co^{3+}OH^-] \qquad (14)$$

$$[Co^{3+}OH^-] + RO_2H \longrightarrow RO_2\cdot + CO^{2+} + H_2O \qquad (15)$$

alcohols and dialkyl peroxides, RO_2R, and RO_2S (where SH represents solvent) (*41*) [Eqs. (14) and (15)].

In the absence of olefins, vanadium (*31*) and molybdenum (*30*) compounds also appear to catalyze this type of reaction, but rather slowly. For example, *t*-butyl hydroperoxide refluxed for 1 hr with molybdenum hexacarbonyl in a benzene–toluene mixture at 87° was only 6.2% decomposed, whereas identical conditions except for the presence of excess 2-octene led to 80% decomposition of the hydroperoxide in 1 hr (*30*).

The metal-catalyzed epoxidation, then, is not a free-radical reaction (*29,30*), although side reactions accounting for perhaps 5–10% of the hydroperoxide decomposition probably are. Instead, there is general agreement (*29,32*) that metal ion–hydroperoxide complexes are formed that either make the peroxidic oxygen atoms more electropositive or stabilize an incipient alkoxide ion, or both, thus providing the driving force for nucelophilic displacement by the electrons of the carbon–carbon double bond [Eqs. (10)–(12)].

$$M(L)_n + RO_2H \underset{}{\overset{\text{fast}}{\rightleftharpoons}} (L)_{n-1}M^{\delta-}\cdots\underset{\mid\atop H}{\overset{O-R\atop\mid}{O}}{}^{\delta+} + L \quad \text{(L is any ligand)} \tag{10}$$

$$(L)_{n-1}M^{\delta-}\cdots O^{\delta+}(H)(OR) + {>}C{=}C{<} \xrightarrow{\text{slow}} (L)_{n-1}M(HOR) + {>}C\overset{O}{—}C{<} \tag{11}$$

$$(L)_{n-1}M(HOR) + RO_2H \overset{\text{fast}}{\rightleftharpoons} (L)_{n-1}M\cdots\underset{\mid\atop H}{\overset{O-R\atop\mid}{O}} + ROH \tag{12}$$

This peracid type of mechanism is supported by the similarity in reactivity (Sections I.C and II.C), the rate-enhancing effect of electron-withdrawing substituents on R (Section II.b), and by kinetic measurements. Although the early work of Indictor and Brill (*22*) seemed to indicate half-order dependence on metal-ion concentration, more recent studies (*30,31*) have shown first-order dependence, at least for vanadium and molybdenum.

Using *t*-butyl hydroperoxide, molybdenum hexacarbonyl, and 1-octene or 2-octene in benzene, Sheng and Zajacek (*30*) showed the reaction to be first-order in each component [Eq. (16)].*

$$\text{Rate epoxidation} = k[\text{Mo(CO}_6)]]t\text{—BuO}_2\text{H}]]\text{octene}] \tag{16}$$

Plots of log (*t*-butyl hydroxide) versus time were linear when olefin was present in large excess; the pseudo-first-order rate constant with 1-octene

* It is perhaps worth noting, however, that the order in catalyst concentration appears to rest on only two determinations.

and 3.8×10^{-5} M molybdenum hexacarbonyl at 85.5° was 0.78×10^{-4} sec^{-1}. Pressuring the system with carbon monoxide retarded the reaction, as would be expected for a mechanism involving ligand coordination [Eq. (14)].

Masio and Kato (*29*) have published very detailed rate studies for both *t*-butyl hydroperoxide and cumyl hydroperoxide and molybdenum trioxide and cyclohexene in dioxane and cumene. They found that the rate of epoxidation was proportional to hydroperoxide and olefin concentrations, but could not determine a reaction order for their insoluble catalyst. This factor unfortunately makes their measured activation energies, and differences in reactivity in the two solvents, of dubious value.

Kinetics for vanadyl acetylacetonate–*t*-butyl hydroperoxide in cyclohexene–cyclohexane mixtures (*31*), though first order in metal ion and in olefin, were somewhat more complex, apparently because of a rather low stability constant [K, Eq. (10)] for the vanadium–hydroperoxide complex. The rate expression in neat cyclohexene thus took the form shown in Eq. (17).

$$-\delta RO_2H/\delta t = \frac{k[\text{VO acac}_2]}{(1/[RO_2H]K) + 1} \tag{17}$$

At 24.1° and 50.6°, k and K (*17*) were respectively 7.3 and 39.3 min^{-1}, and 15.3 and 10.3 M^{-1}. For k, $\Delta H^{\neq} = 12.7$ kcal and $\Delta S^{\neq} = -19.8$ e.u.

The kinetic measurements were complicated even more by *t*-butyl alcohol produced in the course of the reaction. This caused severe retardation, probably by competition with *t*-butyl hydroperoxide for complexing sites. Plots of log (*t*-butyl hydroperoxide) versus time were not linear and had to be extrapolated to zero time to provide rate constants.

An alternative mechanism pictures the hydroperoxide merely as a source of hydrogen peroxide from which pervanadic, permolybdic, pertungstic, etc., acids are formed. Equilibration between hydroperoxide and alcohol + hydrogen peroxide [Eq. (18)] can be exceedingly rapid (*42*).

$$t\text{-BuO}_2\text{H} + \text{—}\overset{|}{\underset{|}{\text{V}}}\text{—OH} \rightleftharpoons t\text{-BuOH} + \text{—}\overset{|}{\underset{|}{\text{V}}}\text{—O}_2\text{H} \tag{18}$$

For example, at 60° *t*-butyl hydroperoxide can be titrated by reagents specific for hydrogen peroxide (*43*). Acid catalysis, observed by Gould and coworkers (*31*) in the vanadyl acetylacetonate–*t*-butyl hydroperoxide epoxidation of cyclohexene, would be nicely explained by this scheme.

The mechanism has been rejected by Sheng and Zajacek (*30*) on the grounds that hydrogen peroxide–metal ion and hydroperoxide–metal ion epoxidation show different characteristics. For the former it is alleged that tungsten is the best catalyst, and water or alcohols, the best solvents. For the latter tungsten is characterized as a poor catalyst and molybdenum is claimed to be effective, and nonpolar solvents are preferred. Actually, the argument is not altogether convincing, considering that nothing is known about the effects of catalysts or solvents on the hypothetical reaction in Eq. (19), and even less so in view of Mashio and Kato's findings (*29*) that molybdenum trioxide is a somewhat better catalyst than tungsten trioxide for epoxidation with hydrogen peroxide.

A more significant observation by the latter workers (*29*) was that, while hydrogen peroxide–molybdenum trioxide reagents exhibited a strong absorbance maximum at 337 mμ, indicative of permolybdic acid, *t*-butyl hydroperoxide–molybdenum trioxide mixtures showed no such absorption. Even this does not completely eliminate a peracid mechanism since its steady-state concentration might be very small. It would be useful to have some kinetic data for nontertiary hydroperoxides.

Both Gould et al. (*31*) and Sheng and Zajacek (*30*) allude to the need for catalyst "activation" prior to reaction. The basis for Gould's comments were rapid transient color changes on mixing *t*-butyl hydroperoxide with the vanadium catalysts. Sheng and Zajacek do not say, but one suspects they may have observed induction periods. The nature of the activated species is said to be under investigation (*31*).

There has been virtually no discussion of *why* molybdenum, vanadium, and tungsten are the best catalysts for hydroperoxide epoxidation, although, as observed by Gould (*32*), molybdenum(VI), tungsten(VI), and vanadium (V) are similar in being small, highly charged ions having vacant, low-lying orbitals.

A hydroperoxide–catalyst complex has also been proposed by Wolf and Barnes (*39*) for the epoxidations promoted by alkyl metaborates. In the absence of olefin, the reaction appears to take an ionic course rather than one involving free radicals. Thus, cumene hydroperoxide yields acetone and phenol rather than acetophenone.

Note added in proof

The balance between journal and patent literature is gradually being redressed. The list of olefins for which published yields are available (Section II.C) has been augmented (N. M. Sheng and J. G. Zajacek, *J. Org. Chem.*, **35,** 1839 (1970) and now includes many bifunctional compounds. The results furnish additional qualitative evidence for steric and polar influences on selectivity, and for the general superiority of molybdenum over vanadium catalysts (Section II.A and E). Epoxidation of allylic alcohols is an exception to the latter, the stronger complexation of alcohols by vanadium apparently having a synergistic effect here. (In a reaction related to epoxidation, the oxidation of triphenylphosphine to its oxide by hydroperoxides, rate studies have shown molybdenum to be 10 times more effective than vanadium as a catalyst (R. Hiatt and C. McColeman, *Can. J. Chem.*, in press).)

Quantitative comparisons of polar influences on selectivity (Section II.C) of peracids vs. t-BuO$_2$H-molybdenum naphthenate are now extant. Hammett ρ-values fall between -1.4 and -1.7 for both reagents, not only for epoxidation of substituted styrenes (yields quantitative at 60°C) (G. R. Howe and R. Hiatt, *J. Org. Chem.*, in press), but for the oxidation of substituted anilines to nitro aromatics as well (G. R. Howe and R. Hiatt, *J. Org. Chem.*, **35**, 4007 (1970)).

It is now apparent that even for molybdenum catalysis, simple kinetics, first order in each of the three components (Section IV) are the exception. (N. M. Sheng, J. G. Zajacek, and T. N. Baker III, Reprints, Symposium on New Olefin Chemistry, Division of Petroleum Chemistry, The American Chemical Society, Houston, Texas, February, 1970; G. R. Howe and R. Hiatt, *J. Org. Chem.*, in press.) These findings detract little from basic conclusions about the mechanism, but illustrate complexities yet to be fully understood.

REFERENCES

1a. N. A. Milas, *J. Amer. Chem. Soc.*, **59**, 2342 (1937).
1b. N. A. Milas and S. Sussman, *J. Amer. Chem. Soc.*, **58**, 1302 (1936).
2. W. Treibs, *Chem. Ber.*, **72B**, 7 (1939)
3. M. Mugden and D. P. Young, *J. Chem. Soc.*, 2988 (1949).
4. J. M. Church and R. Blumberg, *Ind. Eng. Chem.*, **43**, 1780 (1951).
5. R. P. Linstead, L. N. Owen, and R. F. Webb, *J. Chem. Soc.*, 1218 (1953).
6. Machu, *Wasserstoflperoxyd and die Perverbindungen*, Springer Verlag, Wien, 1937, p. 240.
7. British Patent 837,464 (1957) (*Chem. Abstr.*, **55**, 566).
8a. G. B. Payne and P. H. Williams, *J. Org. Chem.*, **24**, 54 (1959).
8b. U.S. Patent 2,786,854 (1957) (*Chem. Abstr.*, **51**, 14791).
8c. U.S. Patent 2,833,787 (1958) (*Chem. Abstr.*, **52**, 16367).
8d. U.S. Patent 2,833,788 (1958) (*Chem. Abstr.*, **52**, 16367).
9. P. G. Sergeev and L. M. Bukneeva, *Zh. Obshch. Khim.*, **28**, 101 (1958). (*Chem. Abstr.*, **52**, 12758).
10. Z. Raciszewski, *J. Amer. Chem. Soc.*, **82**, 1267 (1960).

11. H. C. Stevens and A. J. Kaman, *J. Amer. Chem. Soc.*, **87,** 734 (1965).
12. N. C. Yang and R. A. Finnegan, *J. Amer. Chem. Soc.*, **80,** 5845 (1958).
13. J. Q. Payne, *J. Org. Chem.*, **24,** 2048 (1959); **25,** 275 (1960).
14. U.S. Patent 2,508,256.
15. D. E. VanSickle, F. R. Mayo, and R. M. Arluck, *J. Amer. Chem. Soc.*, **87,** 4824, 4834 (1965).
16. D. E. VanSickle, F. R. Mayo, R. M. Arluck, and M. Syz, *J. Amer. Chem. Soc.*, **89,** 967 (1967).
17. W. F. Brill, *J. Amer. Chem. Soc.*, **85,** 141 (1963).
18. W. F. Brill and B. J. Barone, *J. Org. Chem.*, **29,** 140 (1964).
19. W. F. Brill, *Selective Oxidation Processes*, Advances in Chemistry Series No. 51, American Chemical Society, Washington, D.C., 1965, pp.70–80.
20. C. Walling and L. Heaton, *J. Amer. Chem. Soc.*, **87,** 48 (1965).
21. E. G. E. Hawkins, *J. Chem. Soc.*, 2169 (1950).
22. N. Indictor and W. F. Brill, *J. Org. Chem.*, **30,** 2074 (1965).
23. J. Kollar and coworkers to Halcon International:
  a. Neth. Appl. 6,507,189 (1965) (*Chem. Abstr.*, **65,** 2222).
  b. Belgian Patent 641,452 (1963).
  c. Belgian Patent 657,838 (1964).
  d. Belgian Patent 680,852 (1966).
  e. U.S. Patent 3,350,422 (1967) (*Chem. Abstr.*, **68,** 2922).
  f. U.S. Patent 3,351,635 (1967) (*Chem. Abstr.*, **68,** 21821).
  g. U.S. Patent 3,360,584 (1967).
  h. British Patent 1,122,731 (1968) (*Chem. Abstr.*, **69,** 96235).
24. K. A. Allison and coworkers to British Petroleum:
  a. Belgian Patent 640,202 (1964).
  b. Belgian Patent 640,204 (1964).
25. N. M. Sheng and J. G. Zajacek to Atlantic Refining:
  a. Canadian Patent 799,502 (1968).
  b. Canadian Patent 799,503 (1968).
  c. Canadian Patent 799,504 (1968).
  d. British Patent 1,119,476 (*Chem. Abstr.*, **69,** 99891).
26. I. S. DeRoch and P. Menguy:
  a. French Patent 1,505,337 (1967) (*Chem. Abstr.*, **69,** 96439).
  b. French Patent 1,505,332 (1967) (*Chem. Abstr.*, **69,** 96440).
27. K. A. Allison, P. Johnson, G. Foster, and M. Sprake, *Ind. Eng. Chem. Prod. Res. Develop.*, **5,** 166 (1966).
28. F. List and L. Kuhnen, *Erdol and Kohle*, **20,** 192 (1967).
29. F. Mashio and S. Kato, *Mem. Fac. Ind. Arts., Kyoto Tech. Univ., 1967*, No. 16, 79; *Yuki Gosei Kagaku Shi 26*, 367 (1968) (*Chem. Abstr.*, **69,** 18476).
30. N. M. Sheng and J. G. Zajacek, Advances in Chemistry Series No. 76, American Chemical Society, Washington, D.C. (1968).
31. E. S. Gould, R. Hiatt, and K. C. Irwin, *J. Amer. Chem. Soc.*, **90,** 4573 (1968).
32. E. S. Gould and M. Rado, *J. Catal.*, **13,** 238 (1969).
33a. J. Rouchaud and P. Nsumba, *Bull. Soc. Chim. Fr.*, **75** (1969).
33b. J. Rouchaud and J. Fripiat, *Bull. Soc. Chim. Fr.*, **78** (1969).
33c. J. Rouchaud and J. Mawanka, *Bull. Soc. Chim. Fr.*, **85** (1969).
34. L. Kuhnen, *Chem. Ber.*, **99,** 3384 (1966).

35. N. M. Sheng and J. G. Zajacek, *J. Org. Chem.*, **33,** 588 (1968).
36. Belgian Patents 661,500 and 668,811 (1965).
37. L. Kuhnen, *Angew Chem. Int.*, **5,** 893 (1966).
38. U.S. Patent 3,251,888 (1966) (*Chem. Abstr.*, **65,** 2147).
39. P. F. Wolf and R. K. Barnes, *J. Org. Chem.*, **34,** 3441 (1969).
40a. D. Swern, *J. Am. Chem. Soc.*, **69,** 1962 (1947).
40b. D. Swern, *Encyclopedia of Polymer Science and Technology*, Vol. 6, p. 83 (1967).
41. R. Hiatt, T. Mill, and F. R. Mayo, *J. Org. Chem.*, **33** ,1416 (1968), and references therein.
42. D. E. Bissing, C. A. Matuszak, and W. E. McEwen, *J. Amer. Chem. Soc.*, **86,** 3824 (1964).
43. H. Pobiner, *Anal. Chem.*, **33,** 1423 (1961).

4

Metal Ion-Catalyzed Oxidation of Organic Substrates with Peroxides

A. R. DOUMAUX, JR.
UNION CARBIDE CORPORATION
CHEMICALS AND PLASTICS
SOUTH CHARLESTON, WEST VIRGINIA

I. Introduction 141
II. Scope and Mechanism 143
III. Reaction Conditions 158
A. Effect of Solvent 158
B. Choice of Oxidants 159
C. Experimental Conditions 166
Appendix 177
References 182

I. INTRODUCTION

Fenton's reagent, discovered in 1894, is perhaps the oldest metal ion–peroxide oxidation reagent known to organic chemists. This reagent, consisting of a 1 to 1 stoichiometry of ferrous sulfate and hydrogen peroxide, contains the simplest and most readily available peroxide from both a historical and commercial standpoint. Consequently, the system has been thoroughly studied (*1*,*2*). From decomposition studies Haber and Weiss (*3*,*4*) proposed a mechanism for the ferrous ion-catalyzed decomposition of hydrogen peroxide involving the formation of a hydroxyl radical as the reactive intermediate [Eq. (1)]. Oxidations with Fenton's

$$H_2O_2 + Fe(II) \longrightarrow HO\cdot + Fe(III)(OH) \qquad (1)$$

reagent have been adequately reviewed (*1*,*2*).

Early work involving the decomposition of organic peroxides such as "cyclohexanone hydroperoxide" (prepared by the addition of hydrogen peroxide to cyclohexanone) by metal ions gave products that were difficult to obtain by other means (*5*,*6*) [Eq. (2)]. Such findings served to stimulate industrial interest.

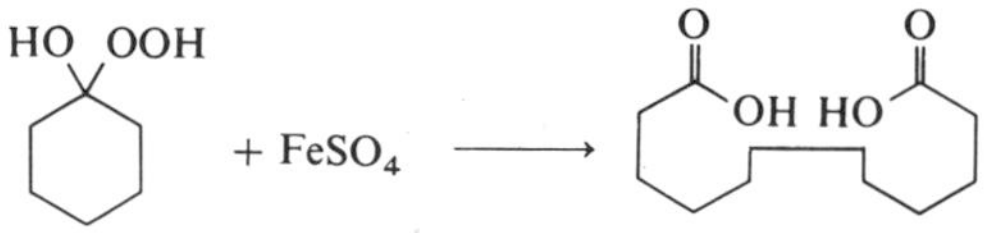

$$\text{(1-hydroxycyclohexyl hydroperoxide)} + FeSO_4 \longrightarrow HOOC(CH_2)_{10}COOH \quad (2)$$

Concurrently, Hawkins (*7–9*) in England and Kharasch et al. (*10–19*) in the United States began to exploit metal ion-catalyzed decomposition of organic peroxides to oxidize other reactive organic substrates present in the reaction mixture. For example, the decomposition of *t*-butylperbenzoate in cumene with cuprous chloride produced α-cumylbenzoate (*13*) [Eq. (3)].

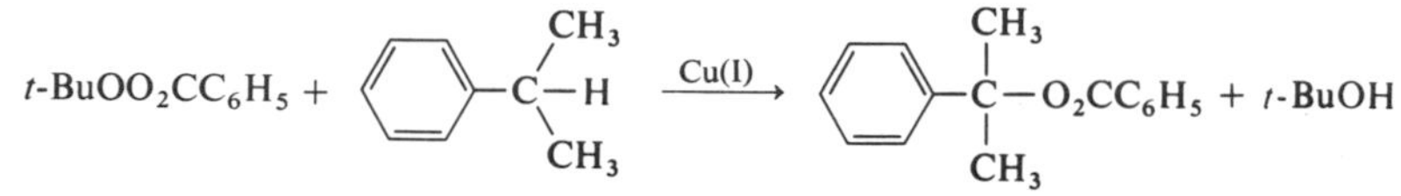

$$t\text{-}BuOO_2CC_6H_5 + C_6H_5C(CH_3)_2H \xrightarrow{Cu(I)} C_6H_5C(CH_3)_2{-}O_2CC_6H_5 + t\text{-}BuOH \quad (3)$$

(3)

In the early 1960s, Sosnovsky and Lawesson (*20*) and Berglund and Lawesson (*21*) further elucidated the scope of this type of oxidation. In general, the results were discouraging and interest in this oxidation method as a synthetic tool faded.

While endeavors for synthetic exploitation lessened, work on the reaction mechanism continued, mainly by Kochi et al. (*22,23*) working, at first, with simple olefins, copper ion, and peresters and later with alkyl radicals. Their work centered on the generation of alkyl radicals from diacylperoxides that were decomposed in the presence of copper ions. Though the endeavors of Kochi et al. have served to elucidate the overall mechanism of this type of oxidation, considerably more investigation will be required before a clear understanding of the catalytic species involved is achieved (*vide infra*). From mechanism studies by Kochi et al., however, and from work by other investigators (*24,25*), it is clear that a reinvestigation of the synthetic utility of this type of oxidation is justified.

A careful review of the literature showed that several types of organic peroxides have been used previously to synthetic advantage. As might be expected, yields and products on oxidation of a particular substrate changed drastically depending on choice of peroxide, metal ion, and reaction conditions (see Section III). Recent results (*24,25*) have shown that with a proper choice of reactants and conditions, novel and selective oxidations can be achieved under mild conditions. With peracetic acid (*25*) and other peracids, oxidations of compounds generally considered inert can be achieved easily.

In the past several years the oxidation and reduction of organic compounds with metal ions and the redox reactions of metal ions alone have received a great amount of study. The reader is referred to several excellent reviews (*26–30*). Much of the knowledge gained in these fields can be applied to the metal ion-catalyzed oxidations of organic substrates. Certainly, the same major problem is involved—defining the role of the metal ion in the reaction.

II. SCOPE AND MECHANISM

In an oxidation-reduction sequence between metal ions in solution, two transition states for electron transfer are possible. To the inorganic chemist (*29,30*) these two transition states are described as "outer-sphere" and "inner-sphere" transition states. In both cases atom or electron transfer is possible. In an outer-sphere transition state, the inner or first coordination shell of the metal ion is substitution-inert and remains intact during the reaction. In an inner-sphere reaction, the first coordination shell of the metal ion is substitution-labile. Substitution must occur first, forming a bridging group common to both metal ions. In the past this has been referred to as the "bridged activated complex" (*30*). Atom transfer has been used to describe oxidations that proceed with the transfer of an atom or group from oxidant to reductant (or vice versa). The atom or group transferred may or may not be electrically neutral. For example, transfer of OH^+ or OH^- would be considered as atom transfers.

Examples of the four possible transition states between metal ions are shown below. Where possible, analogous reactions between metal ions and organic radicals are shown. The electron transfer between $Fe(CN)_6^{4-}$ and $Fe(CN)_6^{3-}$, both substitution-inert, has been shown to proceed by an outer-sphere mechanism. This means the electron transferred has to pass through the first coordination shell to reduce the metal ion. The reaction of mercaptans in the presence of excess CN^- with ferricyanide ion can be considered the organic equivalent (*31*). Electron transfer from the mercapto anion must occur through the substitution-inert shell of the ferricyanide ion [Eqs. (4) and (5)].

$$RS^- + Fe(CN)_6^{3-} \longrightarrow RS\cdot + Fe(CN)_6^{4-} \quad (4)$$

$$2RS\cdot \longrightarrow RSSR \quad (5)$$

An example of atom transfer taking place via an outer-sphere mechanism has been suggested (*32*) for the Fe^{2+}–$FeOH^{2+}$ exchanges. The transition state may be depicted as follows (**1**) (*29*).

$$(H_2O)_5Fe(II)\text{—}\overset{\displaystyle H}{\overset{|}{O}}\cdots H\text{—}\underset{\underset{\displaystyle H}{|}}{O}\text{—}Fe(III)(H_2O)_5$$

(**1**)

Atom transfer between metal ions has been well documented by the elegant work of Basolo and Pearson (*29*) and Taube (*30*). Using a series of $Co(III)(NH_3)_5L$ compounds as oxidants and Cr^{2+} as the reductant, Taube has shown that the transfer of the ligand (L) occurs for L = Br^-, Cl^-, NCS^-, N_3^-, acetate, oxalate, SO_4^{2-} and others. For this particular combination of metal ions the transfer of L is fastest for the halogens [Eq. (6)].

$$Cr(II) + Co(III)(NH_3)_5L \longrightarrow Cr(III)L + Co(II) + 5NH_3 \tag{6}$$

The oxidation of acetate ion by Co(III) might be considered an organic counterpart (*33*) [Eqs. (7) and (8)]. The reactions of alkyl radicals and

$$Co(III)(H_2O)_6 + CH_3CO_2H \rightleftharpoons (H_2O)_5Co(III)(O_2CCH_3) + H_2O + H^+ \tag{7}$$

$$H_2O + (H_2O)_5Co(III)(O_2CCH_3) \longrightarrow (H_2O)_6Co(II) + CH_3CO_2\cdot \tag{8}$$

several cupric or ferric salts can also be considered as atom transfer reactions (*34–36*). For example, Minisci and Galli (*34,35*) have reported the formation of azides by using ferrous sulfate, sodium azide, ferric chloride, cyclohexene, and hydrogen peroxide via Eqs. (9)–(12).

$$Fe(II) + HOOH \longrightarrow Fe(III)(OH) + HO\cdot \tag{9}$$

$$HO\cdot + Fe(II)N_3 \longrightarrow Fe(II)(OH) + N_3\cdot$$

$$N_3\cdot + \text{cyclohexene} \longrightarrow \text{2-azidocyclohexyl radical} \tag{10}$$

$$\text{2-azidocyclohexyl radical} + Fe(II)Cl \longrightarrow \text{1-azido-2-chlorocyclohexane} + Fe(II) \tag{11, 12}$$

In this situation the transferred group acts as a free radical. Thus in the transition state little change develops. Taube (*27,29*) has shown that electron transfer can also occur through an extended bridge of atoms. For example, Cr(II) can be oxidized by pentamine Co(III) carboxylates [Eq. (13)].

$$\text{Co(III)}-\text{O}_2\text{CR} + \text{Cr(II)} \longrightarrow [\text{Co}\cdots\text{O}-\text{C(R)}=\text{O}\cdots\text{Cr}] \longrightarrow \text{Co(II)} + \text{RCO}_2-\text{Cr(III)} \quad (13)$$

Sebera and Taube (*37*) and Fraser and Taube (*38*) have also shown that electron transfer can occur through extensively conjugated systems such as fumarate and terephthalate ions.

An example of electron transfer through a bridged complex which is not accompanied by atom transfer has been suggested by Halpern and Candlin (*39*) in the reaction between formatopentamine Co(III) and permanganate ion [Eq. (14)].

$$[(\text{NH}_3)_5\text{Co(III)OCHO}]^{2+} + \text{MnO}_4^- \longrightarrow [\text{Co(III)(NH}_3)_5(\text{CO}_2^- \cdot)]^{2+} + \text{HMnO}_4^- \rightarrow \text{Co(II)} + \text{CO}_2 \quad (14)$$

The present knowledge of the role of the metal ion in the oxidation-reduction of organic substrates is severely limited. The study of the metal ion in such redox processes should include (a) determination of whether an inner- or outer-sphere transition state is involved, (b) isolation or detection of intermediates, if possible, and (c) determination of the composition and geometry of the intermediate and/or transition states involved. At the present time a clear choice of whether an outer- or inner-sphere transition state is involved is offered only in certain specific cases. Some information concerning the formation of metal–hydroperoxy radical complexes between transition metal ions and peroxides has been established by esr experiments (*40–41*). One example of an isolable copper–hydrogen peroxide complex has been reported (*42*). In solution the degree of aggregation of the metal ion may not be known or can change as the reaction proceeds. Also, ligand exchange processes can occur, changing the first coordination shell around the metal ion. This latter problem is especially pertinent because the ligands in the first coordination shell surrounding a metal ion control (a) the geometry of the metal–ion complex, (b) the electronic structure and hence the redox potential of the couple, (c) the probability of barrier penetration (for an outer-sphere mechanism), (d) the degree of agglomeration of the metal ions, and (e) the strength with which competitive ligands are bound to the metal ion. The theories concerning the role of ligands on metal ions are beyond the scope of this chapter, and excellent reviews on the subject are available (*29,30*).

Investigation of the mechanism of the metal ion-catalyzed oxidation with peroxides has been largely confined to the oxidation of alkyl radicals produced from acyl peroxides in the presence of copper salts. The following discussion is based largely on these findings.

In the presence of an organic substrate with a reactive hydrogen the decomposition of an organic peroxide by Cu(I) can be represented by the general scheme outlined in Eqs. (15)–(17).

$$\mathrm{Cu(I)X} \text{---} \left| \begin{array}{lcl} \mathrm{HOOH} & \longrightarrow & \mathrm{Cu(II)(OH)X + HO\cdot} \\ \mathrm{ROOH} & \longrightarrow & \mathrm{Cu(II)(OH)X + RO\cdot} \\ \mathrm{ROOR} & \longrightarrow & \mathrm{Cu(II)(OR)X + RO\cdot} \\ \mathrm{ROO\overset{O}{\overset{\|}{C}}R} & \longrightarrow & \mathrm{Cu(II)(O_2CR)X + RO\cdot} \\ \mathrm{R\overset{O}{\overset{\|}{C}}\text{---}O\text{---}O\overset{O}{\overset{\|}{C}}R} & \longrightarrow & \mathrm{Cu(II)(O\overset{O}{\overset{\|}{C}}R)X + RCO_2\cdot} \\ & & \qquad\qquad \downarrow \\ & & \qquad\qquad \mathrm{R\cdot + CO_2} \end{array} \right. \tag{15}$$

(15a)

$$\left.\begin{array}{l} \mathrm{HO\cdot} \\ \mathrm{RO\cdot} \\ \mathrm{RCO_2\cdot} \\ \mathrm{R\cdot} \end{array}\right\} + \text{---}\overset{|}{\underset{|}{C}}\text{---H} \longrightarrow \text{---}\overset{|}{\underset{|}{C}}\cdot + \left\{\begin{array}{l} \mathrm{HOH} \\ \mathrm{ROH} \\ \mathrm{RCO_2H} \\ \mathrm{RH} \end{array}\right. \tag{16}$$

$$\text{---}\overset{|}{\underset{|}{C}}\cdot + \mathrm{Cu(II)X} \xrightarrow[\text{transfer}]{\text{Ligand}} \text{---}\overset{|}{\underset{|}{C}}\text{---X} + \mathrm{Cu(I)} \tag{17a}$$

$$\begin{array}{l} \text{---}\overset{|}{\underset{|}{C}}\cdot + \mathrm{Cu(II)X} \longrightarrow [\mathrm{R\text{---}CuX}] \\ [\mathrm{R\text{---}CuX}] \xrightarrow[k_e]{\text{Elimination}} \mathrm{R(\text{---}H) + Cu(I) + H^+} \\ \qquad \xrightarrow[k_s]{\text{Substitution}} \mathrm{R'\text{---}S + Cu(I) + H^+} \\ \qquad \text{Solvent HS} \end{array} \tag{17b}$$

Equation (15) indicates the one-electron oxidation of Cu(I) to Cu(II) with concurrent generation of a hydroxy, alkoxy, or acyloxy radical, depending on the peroxide used. Proof for the production of such radical species has been amply demonstrated by Kochi and others by trapping them with conjugated olefins such as butadiene or styrene (*43*). Walling and Zavitsas (*44*) have shown that the relative reactivities of seven olefins were the same with *t*-butylperacetate or *t*-butylperbenzoate in the presence of copper ion as chlorination with *t*-butylhypochlorite. This indicated the *t*-butoxy radical was the common intermediate in both reactions. Acyloxy radical decomposition may occur, producing carbon dioxide and generating a new radical specie [Eq. (15a)]. This process is extremely rapid, where R is alkyl but proceeds much more slowly when R is an aryl group (see Section III.B.2). A second mode of decomposition is important when alkyl hydroperoxides are employed. With Cu(II), for example, a hydroperoxide can be oxidized to an alkyl hydroperoxy radical as shown in Eq. (18). In several instances the sole product isolated is the mixed per-

$$\mathrm{Cu(II)} + \mathrm{ROOH} \longrightarrow \mathrm{Cu(I)} + \mathrm{H^+} + \mathrm{ROO\cdot} \tag{18}$$

oxide. For example, cycloheptane on reaction with *t*-butylhydroperoxide and Cu(I) (*24*) gives the mixed peroxide and cycloheptene [Eq. (19)]. The oxygen atoms in the Cu(II) carboxylate derivative formed in Eq. (15)

$$\text{cycloheptane} + t\text{-BuOOH} + \mathrm{Cu(I)Cl} \longrightarrow \text{cycloheptyl-OOtBu} + \text{cycloheptene} \tag{19}$$

became completely equilibrated, as one might expect. Denney et al. (*45*), using *t*-butylperbenzoate–carbonyl ^{18}O in the presence of cyclohexene and Cu(II), isolated cyclohexenylbenzoate. Reduction (H_2/Pd) gave cyclohexanol and benzyl alcohol. Analysis for ^{18}O indicated complete equilibration between the two oxygens of the cyclohexenylbenzoate.

Equation (16) shows the abstraction of a reactive hydrogen atom from an organic substrate via the alkyl, alkoxy, aryl, aryloxy, or hydroxy radical generated in Eq. (15). Again, evidence for the generation of such radicals has been provided indirectly. The copper-catalyzed oxidation of optically active 2-phenylbutane with *t*-butylperbenzoate gave optically inactive 2-phenylbutan-2-ol on reduction of the reaction product with lithium aluminum hydride (*46*) [Eq. (20)]. A radical intermediate would lead to loss of optical activity. Optically inactive ketone was obtained from the copper-catalyzed oxidation of optically active Δ′-*p*-menthene with

$$\mathrm{C_6H_5{-}\overset{\displaystyle CH_3}{\underset{\displaystyle H^*}{C}}{-}Et} + \mathrm{C_6H_5\overset{\displaystyle O}{\overset{\|}{C}}OOtBu} \xrightarrow[\text{2. LiAlH}_4]{\text{1. Cu(I)/Cu(II)}} \mathrm{C_6H_5{-}\overset{\displaystyle CH_3}{\underset{\displaystyle OH}{C}}{-}Et} \tag{20}$$

t-butylperacetate, hydrolysis of the acetate mixture, and subsequent oxidation (*47*) [Eq. (21)]. Goering and Mayer (*48*) oxidized optically active bicyclo[3.2.1]oct-2-ene. The bicyclo[3.2.1]oct-3-en-2-ol isolated

$$+ \ t\text{-BuOOCCH}_3 \xrightarrow[\text{2. Hydrolysis; 3. (O)}]{\text{1. Cu(I)}} \qquad (21)$$

after saponification was optically inactive and was greater than 99% *exo* [Eq. (22)]. Under the reaction conditions, recovered olefin was still optically pure and the benzoate ester was stable.

$$\text{Opt. act.} \ + \ t\text{-BuOOCC}_6\text{H}_5 \xrightarrow[\text{80°, 30 min; 2. Saponification}]{\text{1. Cu(I)Br, C}_6\text{H}_6} \text{Opt. inact.} \qquad (22)$$

Cross and Whitham (*49*) reported that an axially substituted ester was also obtained in the oxidation of 1-methylene-4-*t*-butylcyclohexane, as shown in Eq. (23). Further, Kochi et al. (*22,23*) have trapped alkyl radicals produced by metal ion-catalyzed decomposition of acylperoxides by the addition of styrene or butadiene [Eq. (24)]. The reactivity of the hydrogen in the abstraction step is of great practical importance. If a nominally

$$+ \ t\text{-BuOO}_2\text{CC}_6\text{H}_5 \xrightarrow[\text{CuCl; 2. Saponification}]{\text{1. C}_6\text{H}_6\text{, 80°, 8 hr}} \qquad (23)$$

$$(\square\text{-CO}_2)_2 + \text{Cu(I)} + \text{butadiene} \xrightarrow[\text{CH}_3\text{CN}]{\text{HOAc}} \text{O}_2\text{CCH}_3 \qquad (24)$$

"unreactive" substrate is utilized, much of the oxidizing power of the peroxide will be lost by unproductive decomposition, and overall efficiency to product based on peroxide will be necessarily low (*5,6,43*).

Equation (17a) involves the reduction of Cu(II) regenerating a Cu(I) specie and the atom transfer of the group X to the reducing specie R·. This may involve an inner- or outer-sphere process, but the inner-sphere route is usually considered dominant. In an atom-transfer process involving radicals such as the neopentyl radical, no rearrangement should occur, i.e., in the transition state no carbonium ion character develops. Kochi and Bemis (*50*) mention such results with neopentyl radicals and Cu(II). Radicals formed by way of Eq. (16) with *alpha* electron-withdrawing groups attached will not undergo an electron-transfer process but will rapidly undergo an atom-transfer process. *alpha*-Cyanoisopropyl radicals are readily oxidized by Cu(II)Cl to *alpha*-chloroisobutyronitrile (**2**) in an inner-sphere atom-transfer reaction (*51*). The use of cupric acetate, however, gives only products derived from radical dimerization (*51*) [Eq. (25)].

$$(H_3C)_2\dot{C}{-}CN \xrightarrow{Cu(II)Cl_2} \underset{\substack{54\% \\ (\mathbf{2})}}{(H_3C)_2C(Cl)CN} + \underset{\substack{20\% \\ (\mathbf{3})}}{H_3C{-}C(CH_3)(CN){-}C(CH_3)(CN){-}CH_3} + \underset{\substack{23\% \\ (\mathbf{4})}}{(H_3C)_2CH{-}C({=}O){-}NH{-}C(CH_3)_2{-}C({=}O){-}NH_2} \qquad (25)$$

$$\xrightarrow{CuII(OAc)_2} 43\%\ \mathbf{3} + 47\%\ \mathbf{4}$$

Thus, when choosing a metal ion catalyst, the ionization potential of the organic radical should be carefully considered. In the above case, carbonium ion formation is not favored due to the *alpha* electron-withdrawing group. Oxidation was achieved readily by an inner-sphere atom-transfer process which did not involve a highly charged transition state. Thus, a proper choice of the anion associated with the metal ion catalyst is also important. The transition state for this process can be represented as a direct transfer of an atom or radical to the carbon radical. Little development of charge occurs, as shown in Eq. (26). While Eqs.

$$R\cdot + Cu(II)X \longrightarrow [R\cdot X{-}Cu(II) \longleftrightarrow R{-}XCu(I)] \longrightarrow R{-}X + Cu(I) \qquad (26)$$

(15), (16), and (17a) have been widely accepted, Eq. (17b) presented in the general mechanism has been only recently proposed (*22,23*). Kochi has elegantly shown by kinetic and deuterium labeling studies that an organo-copper specie is formed in some cases. Olefin formation from the *beta*-arylethyl radicals, to be discussed presently, is an example of an inner-sphere atom-transfer reaction. The other reaction in Eq. (17b) represents an inner-sphere electron-transfer process leading to a carbonium ion specie which in turn can give solvolysis products or new olefins. Rearrangement can certainly occur. For example, Kochi and Bemis (*22*) have shown that cyclobutyl-, allylcarbinyl- and cyclopropylmethyl radicals generated from the corresponding diacyl peroxides all gave the same mixture of products, indicating either a large amount of carbonium-ion character in the transition state or actual generation of such species in solution [Eq. (27)]. It cannot be ascertained whether this is an inner-sphere electron-transfer process with generation of the solvated carbonium

$$\left.\begin{matrix}\text{cyclobutyl}\cdot \\ \text{cyclopropylmethyl}\cdot \\ \text{allylcarbinyl}\cdot\end{matrix}\right\} + \xrightarrow[0^\circ]{Cu(II)/HOAc-CH_3CN} \text{cyclobutyl-OAc} + \text{cyclopropylmethyl-OAc} + \text{allylcarbinyl-OAc} \quad (27)$$

ion or whether the carbonium ion specie formed within the first coordination shell reacts with another ligand bound to the metal (or exists as an ion pair within a solvent cage).

In a series of experiments Kochi has shown that *beta*-arylethyl radicals produced by decomposition of the corresponding diacyl peroxides led to both elimination and substitution products. The overall sequence as proposed by Kochi et al. (*22,23*) is presented in Eqs. (28)–(33).

$$(ArCH_2CH_2CO_2)_2 + Cu(I)X \xrightarrow[0^\circ]{HOAc-CH_3CN} ArCH_2CH_2CO_2\cdot + XCu(II)(O_2CCH_2CH_2Ar) \quad (28)$$

$$ArCH_2CH_2CO_2Cu(II)X + HOAc \longrightarrow ArCH_2CH_2CO_2H + Cu(II)(OAc)X \quad (29)$$

$$ArCH_2CH_2CO_2\cdot \longrightarrow CO_2 + ArCH_2CH_2\cdot \quad (30)$$

$$ArCH_2CH_2\cdot + Cu(II) \rightleftharpoons [ArCH_2CH_2Cu] \quad (31)$$

$$[ArCH_2CH_2Cu] \xrightarrow{k_e} ArCH{=}CH_2 + Cu(I) + H^+ \quad (32)$$

$$[ArCH_2CH_2Cu] \xrightarrow[HOAc]{k_s} ArCH_2CH_2OAc + Cu(I) + H^+ \quad (33)$$

Table I shows the effect of various ring substituents on the yield and rates of formation of the two products. As the concentration of styrene, the elimination product, increases, attack by the *beta*-phenylethyl radical occurs giving rise to secondary reaction products. Again oxidation by Cu(II) leads to elimination and substitution products [Eqs. (34)–(36)]. Kochi corrected the data presented in Table I by taking these second-generation products into consideration.

$$ArCH_2CH_2\cdot + ArCH{=}CH_2 \longrightarrow ArCH_2CH_2CH_2\dot{C}HAr \quad (34)$$

$$ArCH_2CH_2CH_2\dot{C}HAr \xrightarrow{k_e} ArCH_2CH_2CH{=}CHAr + Cu(I) + H^+ \quad (35)$$

$$ArCH_2CH_2CH_2\dot{C}HAr \xrightarrow[HOAc]{k_s} ArCH_2CH_2CH_2CH(OAc)Ar + Cu(I) + H^+ \quad (36)$$

The data in Table I clearly show that as substituents which can stabilize carbonium ions (p-MeO > p-Me > H > m-MeO) are placed on the aromatic ring, the yield of acetate increases at the expense of the elimination product. Oxidation of 1,1-dideutero-2-(4′-methoxyphenyl)ethyl radicals by Cu(II) yielded an equimolar mixture 1,1- and 2,2-dideutero-2-(4′-methoxyphenyl)ethyl acetate as shown in Eq. (37). Under the reaction conditions employed the 1,1-dideutero-2-(4′-methoxyphenyl)ethyl radical does not

$$CH_3O{-}C_6H_4{-}CH_2CD_2^{\cdot} + Cu(II) \xrightarrow{k_s} H_3C{-}O{-}C_6H_4{-}CH_2CD_2OAc + CH_3O{-}C_6H_4{-}CD_2CH_2OAc \quad 50\%$$

$$CH_3O{-}C_6H_4{-}CH_2CD_2^{\cdot} + Cu(II) \xrightarrow{k_e} CH_3O{-}C_6H_4{-}CH{=}CD_2 \quad 50\%$$

(37)

rearrange. Further, no substitution products arising from a 1,2-hydride shift were found (i.e., no 1-(4′-methoxyphenyl)ethyl acetate. Oxidation of the 2,2-dideutero-2-phenylethyl radical with Cu(II) also gives both 1,1- and 2,2-dideutero-2-phenylethyl acetate, again in equimolar amounts [Eq. (38)]. Whether rapidly equilibrating carbonium ions or bridged

TABLE I

Effect of Ring Substituents on the Relative Rates of Oxidative Elimination and Substitution of β-Arylethyl Radicals

Alkyl radical	% Elimination product (styrene)	% Substitution product (acetate)	$k_e \times 10^{-6}$, liters/mole/sec	$k_s \times 10^{-6}$	$k_e + k_s$
m-$CH_3O-C_6H_4-CH_2CH_2\cdot$	97	3	—	—	—
$C_6H_5-CH_2CH_2\cdot$	95	5	1.6	0.050	1.7
p-$H_3C-C_6H_4-CH_2CH_2\cdot$	60	40	1.0	0.81	1.7
p-$CH_3O-C_6H_4-CH_2CH_2\cdot$	1	99	0.021	1.6	1.5

$$C_6H_5\text{-}CD_2CH_2^{\bullet} + Cu(II) \begin{cases} \xrightarrow{k_s} C_6H_5\text{-}CD_2CH_2OAc + C_6H_5\text{-}CH_2CD_2OAc \quad 50\% \\ \xrightarrow{k_e} C_6H_5\text{-}CD{=}CH_2 \quad 50\% \end{cases} \tag{38}$$

phenonium ions are involved is not pertinent in this discussion. The important fact is that solvolysis products are found in which deuterium scrambling has occurred. While the deuterium label was scrambled, as shown in the substitution products, the elimination reaction gave olefin products with no deuterium scrambling [Eqs. (37) and (38)]. This suggests that a discrete carbonium ion is not involved in the elimination process. Kochi et al. (*23*) have proposed the following transition state for the oxidative elimination reaction in acetic acid, an inner-sphere transition state [Eq. (39)]. Consistent with this picture is the small deuterium isotope

$$\underset{\substack{\text{H} \quad \quad \text{AcO}}}{>C{-}C<}\!\!\text{Cu(II)} \longleftrightarrow \underset{\substack{\text{H} \quad \quad \text{AcO}}}{>C{-}\overset{+}{C}<}\!\!\text{Cu(I)} \longleftrightarrow \underset{\substack{\text{AcO} \quad \text{H}^+}}{>C{=}C<}\!\!\text{Cu(I)} \tag{39}$$

effect found by Kochi (*52*). Independent processes involving different intermediates for the formation of both substitution and elimination products were ruled out. The rates of elimination and substitution, while varying widely themselves, gave a remarkably constant overall rate ($k_e + k_s$) pointing instead to a common intermediate (see Table I). Finally, the ratio of substitution products to elimination products is remarkably dependent on the Cu(II) oxidant in some cases (*22*). This tends to rule out carbonium ions as direct precursors to both the elimination and substitution products. The oxidation state of the proposed organo-copper intermediate was not specified (*23*). However, regardless of the oxidation state of the copper intermediate, the products of decomposition were olefin and substitution products. Similar results have been obtained also by the thermal decomposition of alkyl Cu(I) species (*53–56*). In fact, Whitesides et al (*55*) have found that alkyl copper species decompose rapidly at temperatures above 0°. Thus if alkyl Cu(I) species are formed in metal ion-catalyzed decomposition of acyl peroxides, decomposition would be rapid at the temperatures employed in these reactions.

The same dichotomy has been observed by Evnin and Lam (*24*) in the oxidation of cycloheptane with *t*-butylperoxide and *t*-butylhydroperoxide in the presence of benzamide and various copper salts [Eqs. (40) and (41)].

$$\text{cycloheptane} + (t\text{-BuO})_2 + C_6H_5\overset{O}{\overset{\|}{C}}NH_2 \xrightarrow{C_6H_5Cl} \text{cycloheptyl}\!-\!NH\overset{O}{\overset{\|}{C}}C_6H_5 + \text{cycloheptene} \quad (40)$$

Catalyst (mole % based on peroxide)	% Product	
$FeCl_3$ (1.5)	Trace	> 5
$CuSO_4$ (4)	Trace	30
CuCl (1)	28	43
$CuCl(bipyridine)_2$ (0.5)	45	45
$CuCl_2(phen)_1$ (1)	50	25
$CuCl_2(phen)_2$ (1)	68	18
$CuCN(phen)_2$ (1)	45	12

$$\text{cycloheptane} + t\text{-BuOOH} + C_6H_5\overset{O}{\overset{\|}{C}}NH_2 \xrightarrow[Cu(I)]{C_6H_5Cl} \text{cycloheptyl}\!-\!OOtBu + \text{cycloheptene} \quad (41)$$

Catalyst	% Product	
CuCl	20	30
$CuCl_2(phen)_2$	60	23

In some instances one branch of the bifurcation in the mechanism proposed by Kochi is controlled by other factors. For example, the neopentyl radical has no *beta*-hydrogen, and therefore cannot eliminate to form an olefin; consequently rearrangement is observed (*22*) [Eq. (42)]. Interestingly, some of the rearranged products are olefins. Another factor which favors elimination as the route to product is the stability of the incipient

$$CH_3-\underset{CH_3}{\overset{CH_3}{\underset{|}{\overset{|}{C}}}}-CH_2\cdot + Cu(II) \longrightarrow \text{2-methyl-1-butene} + \text{2-methyl-2-butene} + \text{(2-methyl-2-butyl)OAc} + \text{(2-methyl-2-butyl)NHAc} \quad (42)$$

carbonium ions. The oxidation of 6-methyl-2-piperidone with peracetic acid and $Mn(III)(acac)_3$ gives 6-methyl-3,4-dihydro-2-pyridone as the major product. In this example, carbonium ion formation would not be favored due to the electron-withdrawing *alpha*-amido group [Eq. (43)]. If the Kochi mechanism is general and an alkylmanganese specie is formed,

$$\text{6-methyl-2-piperidone} + CH_3CO_3H \xrightarrow[\text{EtOAc}-\text{HOAc}]{\text{Mn(III)}} \text{6-methyl-3,4-dihydro-2-pyridone} \quad (43)$$

this might be interpreted as an inner-sphere, atom-transfer reaction.

Other examples of metal ion-catalyzed peroxide oxidations which seem to involve carbonium ion species have been reported. The oxidation of 6,6-dimethyl-1,3-cyclohexadiene with *t*-butylperacetate in the presence of cupric acetate gave a small yield of *o*-xylene as well as other products (*44*). The rationale for the formation of this product is shown in Eq. (44).

$$\text{6,6-dimethyl-1,3-cyclohexadiene} + t\text{-BuOOC(O)}CH_3 + \text{Cu(II)} \xrightarrow[\text{Cl}-C_6H_5]{70^\circ} \text{o-xylene (6\%)} + \text{t-BuO, OAc adduct (80\%)} + \text{t-BuO, OAc adduct (8\%)} \quad (44)$$

$$\xrightarrow{t\text{-BuO}\cdot} \text{(A)} \xrightarrow{\text{Cu(II)}} \text{(B)} + \text{Cu(I)} \xrightarrow{\sim CH_3} \text{cation} \xrightarrow{-H^+} \text{o-xylene}$$

Abstraction of hydrogen to produce the radical **A** is followed by electron transfer to Cu(II) producing the cation **B**. This, in turn, is followed by rearrangement and loss of hydrogen ion to yield *o*-xylene. Story (*57–59*) oxidized 2-deuterionorbornadiene with *t*-butylperbenzoate with Cu(I)Br as the catalyst. The product, 7-*t*-butoxynorbornadiene, isolated in 25% yield, contained deuterium approximately statistically distributed. On the basis of these results, Story proposed the mechanism shown by Eqs. (45–(47) for the formation of the product. Addition of the *t*-butoxy radical

$$t\text{-BuOO}_2\text{CC}_6\text{H}_5 + \text{Cu(I)} \xrightarrow{\text{benzene}} t\text{-BuO}\cdot + \text{Cu(II)O}_2\text{CC}_6\text{H}_5 \quad (45)$$

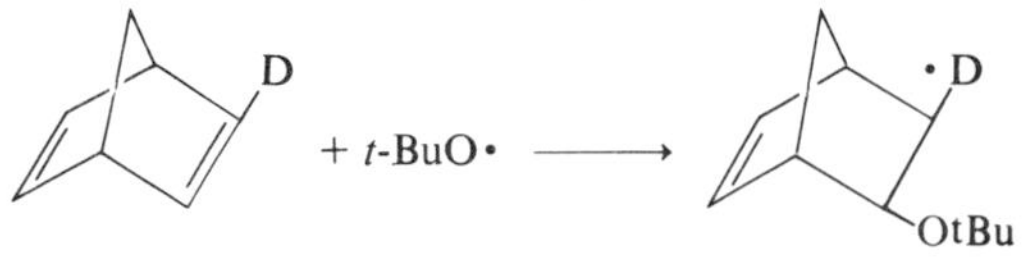

D + *t*-BuO· ⟶ ·D OtBu (46)

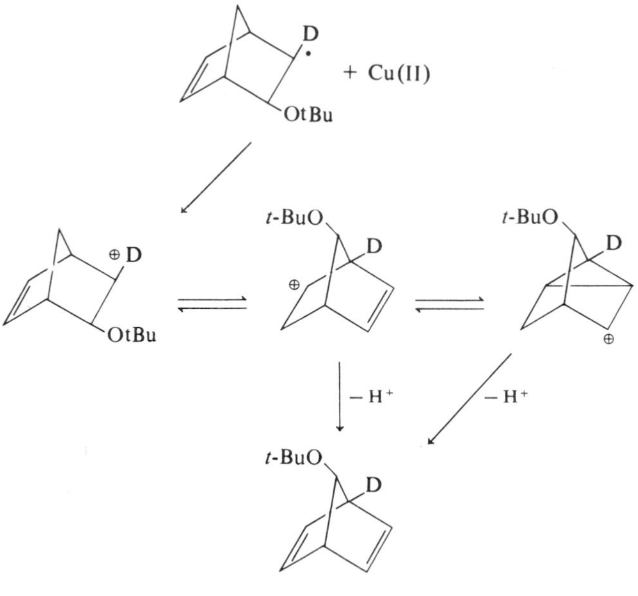

D • + Cu(II) OtBu

⊕D OtBu ⇌ *t*-BuO D ⊕ ⇌ *t*-BuO D ⊕ (47)

$-\text{H}^+$ $-\text{H}^+$

t-BuO D

can occur at the labeled or unlabeled double bond, followed by oxidation to give the radicals **5** and **6**. This oxidation would lead to the observed deuterium scrambling. Oxidation of benzonorbornadiene by Tanida and

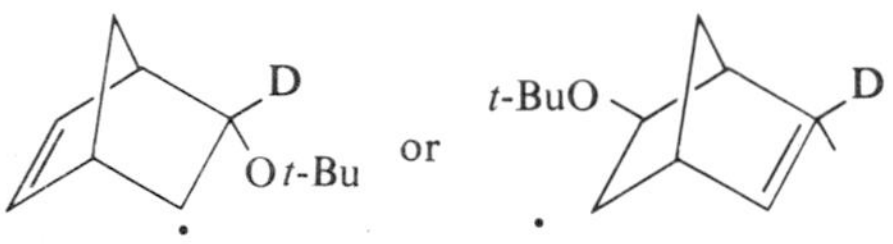

D O*t*-Bu • (**5**) or *t*-BuO D • (**6**)

Tsuji (*60,61*) gave only 7-benzoyloxybenzonorbornadiene (48 %) [Eq. (48)]. Deuterium labeling experiments similar to those of Story's gave equivalent results (*61*).

$$\text{(benzonorbornadiene)} + (C_6H_5\overset{O}{\overset{\|}{C}}{-}O)_2 + CuBr \longrightarrow \text{(7-benzoyloxybenzonorbornadiene, } C_6H_5\overset{O}{\overset{\|}{C}}O{-}) \quad (48)$$

Recently, Evnin and Lam (*24*) reported that toluene reacted with *t*-butylperoxide and di(phenylsulfonyl)amine in the presence of trace amounts of cuprous chloride to give phenyltolylmethanes as the major products [Eq. (49)]. The isomer ratio obtained by Evnin and Lam was very similar to that obtained by the use of benzylation in other systems (*62,63*). The

$$C_6H_5CH_3 + (t\text{-BuO})_2 + (C_6H_5SO_2)_2NH \xrightarrow[\text{CuCl}]{110^\circ,\ 2\ \text{hrs}}$$

$$\underset{21\%}{CH_3C_6H_4{-}CH_2C_6H_5} + \underset{2.3\%}{C_6H_5CH_2Ot\text{-Bu}} + \underset{7.3\%}{C_6H_5CH_2N(SO_2C_6H_5)_2} \quad (49)$$

benzyl *t*-butyl ether as well as the substituted di(phenylsulfonyl)benzylamine isolated from the reaction mixture may also be the products of a benzyl carbonium ion.

While the evidence for carbonium ions in the above cases is quite convincing, the route of their formation (outer-sphere or inner-sphere electron transfer) is unknown. Such may not be the case with the example shown in Eq. (50).

1-Butene or *cis*- and *trans*-2-butene, upon oxidation with *t*-butylperbenzoate and copper salts, give remarkably similar product mixtures of methylallyl(90%)- and crotyl(10%)- benzoates [Eq. (50)]. Both the starting

$$\left.\begin{matrix}\text{1-butene}\\ \textit{cis}\text{-2-butene}\\ \textit{trans}\text{-2-butene}\end{matrix}\right\} + Cu(I\text{–}II) + t\text{-BuOO}_2CC_6H_5 \longrightarrow \underset{90\%}{CH_3CH(O_2CC_6H_5)CH{=}CH_2} + \underset{10\%}{CH_3CH{=}CHCH_2O_2CC_6H_5} \quad (50)$$

olefins and the esters were shown to be stable under the reaction conditions (*64,65*). Solvents such as methanol, *t*-butyl alcohol, and water had little effect on the yield of ester, although a small percentage of ethers and/or alcohols was isolated along with the benzoate esters. When acetic acid was used as the solvent, however, mainly acetates were isolated from the reaction mixture (*65*). Variation of substitution-labile copper complexes had little effect on the ratio of the methylallyl-to-crotyl esters (*66*). When bipyridine or 1,10-phenanthroline was utilized as the ligand, the ratio of acetates (from acetic acid solutions) became almost 50 : 50 (*66*) as the ratio of ligand to copper ion approached 3 : 1. Kochi states (*66*), "The specificity of forming terminally unsaturated derivatives from allylic radical intermediates by simple copper salts is lost by the use of copper phenanthroline or related complexes." He further suggests that solvated allylic carbonium ions are formed, since the distribution obtained is that expected from such species (*66*). The reaction between the allylic radical formed and the bipyridyl copper species (and related species) must be occurring by an outer-sphere electron-transfer process since such coordinating ligands are more substitution-inert. The results with the other copper complexes in which a product ratio of 90/10 was obtained may involve an allyl–copper specie. The results certainly indicate the presence of an intermediate capable of equilibration between the *alpha*-methallyl and *alpha*-crotyl systems. This specie reacts then by an inner-sphere mechanism. In Eq. (50) the ratio between the 3- and 1-substituted butenes is still 3/1 to 9/1 in some solvent mixtures. These results suggest that the same allyl–copper specie is still present and that reaction to products proceeds largely by an inner-sphere process. Alcohols, water and acetic acid might enter the inner coordination shell by rapid metathesis reactions [Eq. (51)].

$$\mathrm{Cu(II)X + HS \rightleftharpoons Cu(II)S + HX} \qquad (51)$$

$$\mathrm{HS = H_2O,\ \ HOAc,\ \ }t\mathrm{\text{-}BuOH,\ \ MeOH}$$

Kochi et al. (*67,68*) have documented several examples of such metathesis in the recent literature. Recently, Cu(I)–olefin complexes have been reported in the literature (*69, 70*). Such species are isolable, well-defined compounds, and could beformed easily during the reaction.

III. REACTION CONDITIONS

A. Effect of Solvent

The choice of the "right" solvent for a metal ion-catalyzed peroxide oxidation is quite important. **First, the solvent must be fairly inert to oxidation relative to the reactive substrate. Benzene, chlorobenzene, aliphatic**

hydrocarbons, nitromethane, acetonitrile, ethyl acetate, *t*-butyl alcohol, tetramethylene sulfone, water, acetic acid, or formic acid meet this first criterion. The use of other solvents, such as tetrahydrofuran, acetone, methanol, and ethyl ether, has been reported. These solvents, especially the ethers, react rapidly with *t*-butylhydroperoxide and violently with peracetic acid. Thus, the use of such solvents leads to a variety of unwanted side reactions.

Second, a solvent which permits homogeneous reaction conditions including solubilization of the metal-ion catalyst is desirable. In this respect benzene, chlorobenzene, and aliphatic hydrocarbons are poor choices, since most metal salts have only limited solubility in these solvents. However, decomposition of the peroxidic materials commonly lead to alcohols, acids, or water, and this in itself tends to ameliorate the problem. Also, mixed solvent systems can be used. Benzene, ethyl acetate, acetonitrile, diethyl ether, tetramethylene sulfone, *t*-butyl alcohol, and acetone seem to have little effect on the yield of product or isomer distribution. Further, the use of water, methanol, formic acid, or nitrogenous compounds such as pyridine, 2,6-lutidine, and 2,4,6-collidine have a pronounced effect on the isomer ratio and/or yield of product. These effects can be directly attributed to the interaction of the solvent and the metal ion such as, in this case Cu(II), by metathesis reactions discussed earlier.

B. Choice of Oxidants

1. Alkyl Hydroperoxides

Of the alkyl hydroperoxides, cumene hydroperoxide and *t*-butylhydroperoxide are most commonly used. Both hydroperoxides are commercially available (see Appendix). Of these two, *t*-butylhydroperoxide is favored, the major by-product being *t*-butyl alcohol, which is inert and in most cases easily removed from the reaction mixture. Cumene hydroperoxide, on the other hand, gives 2-phenyl-2-propanol and *alpha*-methylstyrene as major by-products, both of which can interfere with the isolation of desired products. Both the *t*-butoxy and *alpha*-cumyl radicals formed on oxidation by the metal ion are relatively stable to unimolecular decomposition at the low reaction temperatures generally employed [Eq. (52)]. Tertiary

$$R{-}C(CH_3)_2{-}O\cdot \longrightarrow R{-}C(=O){-}CH_3 + CH_3\cdot \qquad (52)$$

$R = CH_3^-, C_6H_5^-$

alkoxy radicals higher than *t*-butoxy (produced from hydroperoxides, peresters, diacyl peroxides, or peroxides) more readily cleave to ketone and an alkyl radical (*44,67,71,73*). The alkyl radical produced is oxidized rapidly by Cu(II) even in the presence of reactive hydrogen donors (*72*) [Eqs. (53) and (54)]. Therefore, the use of tertiary hydroperoxides (or peroxides and peresters) containing tertiary alkyl groups other than *t*-butyl

$$RCH_2CH_2-\underset{CH_3}{\overset{CH_3}{\underset{|}{\overset{|}{C}}}}-O\cdot \longrightarrow RCH_2CH_2\cdot + CH_3\overset{O}{\overset{\|}{C}}CH_3 \tag{53}$$

$$RCH_2CH_2\cdot + Cu(II) \longrightarrow RCH{=}CH_2 + Cu(I) + H^+ \tag{54}$$

or *alpha*-cumyl (or the like) will lead to undesired decomposition products and side reactions involving attack by the different radicals produced. At the same time the efficiency of the peroxide oxidation to the desired product will be greatly decreased.

In the presence of an active hydrogen donor, or radical acceptor, alkyl hydroperoxides can react with metal ions by two different pathways [Eqs. (55) and (56)]. Equation (55) represents the one-electron oxidation

$$ROOH + m^n \longrightarrow RO\cdot + m^{n+1}(OH) \tag{55}$$

$$ROOH + m^n \longrightarrow ROO\cdot + m^{n-1} + H^+ \tag{56}$$

of the metal ion (m) in the *n*th oxidation state and the reduction of the alkyl hydroperoxide. This is identical to Eq. (15). Equation (56) depicts the one-electron reduction of the metal ion with concurrent production of an alkyl hydroperoxy radical. Evidence for the production of alkyl hydroperoxy radicals was reported originally by Kharasch et al. (*11*). *t*-Butylhydroperoxide treated with catalytic amounts of cobalt naphthenate in the presence of butadiene gave di-*t*-butylperoxides as the only isolable products [Eq. (57)]. Kochi (*74*), however, in similar experiments with

$$CH_2{=}CH{-}CH{=}CH_2 + t\text{-BuOOH} \xrightarrow{Co(II)} t\text{-BuOOCH}_2CH{=}CHCH_2OO\,t\text{-Bu} + t\text{-BuOOCH}_2\underset{OO\,t\text{-Bu}}{CH}CH{=}CH_2 \tag{57}$$

$CuSO_4$, *t*-butylhydroperoxide, and butadiene, found no *t*-butylperoxides as products. Whether the *t*-butylhydroperoxy radical is formed as a free entity initially (which Kharasch postulated) or whether metathesis occurred is not known [Eqs. (58)–(61)]. One explanation is that, while the carbo-

$$t\text{-BuOOH} + \text{Cu(I)} \longrightarrow t\text{-BuO}\cdot + \text{Cu(II)(OH)} \quad (58)$$

$$\text{Cu(II)(OH)} + t\text{-BuOOH} \longrightarrow \text{Cu(II)(OO}t\text{-Bu)} + H_2O \quad (59)$$

$$t\text{-BuO}\cdot + \text{RH} \longrightarrow t\text{-BuOH} + \text{R}\cdot \quad (60)$$

$$\text{R}\cdot + \text{Cu(II)O}_2\,t\text{-Bu} \longrightarrow \text{RO}_2\,t\text{-Bu} + \text{Cu(I)} \quad (61)$$

xylate salts are more stable, the atom transfer of a carboxyl radical may not be favored (*51*). The *t*-butylhydroperoxy radical, on the other hand, may undergo a more facile atom-transfer reaction [Eq. (62)]. Oxidation of R• to the carbonium ion specie followed by nucleophilic attack by

$$\begin{array}{ccc} \text{Cu(II)(O}_2\text{CR)} + t\text{-BuOOH} & \rightleftharpoons & \text{Cu(II)OO}t\text{-Bu)} + \text{RCO}_2\text{H} \\ \big\downarrow \text{R}'\cdot\text{slow} & & \big\downarrow \text{R}'\cdot\text{fast} \\ \text{R}'\text{CO}_2\text{R} + \text{Cu(I)} & & \text{R}'\text{OO}t\text{-Bu} + \text{Cu(I)} \end{array} \quad (62)$$

t-butylhydroperoxide is a more attractive alternative. This latter explanation seems to fit many of the experiments where unsymmetrical peroxides have been found. Where copper halides have been used as the catalyst, halogen atom transfer might occur providing a halide intermediate which is then solvolyzed by *t*-butylhydroperoxide.

2. Diacyl Peroxides

Diacyl peroxides have not been widely used in the past for metal ion-catalyzed oxidations. Aliphatic diacyl peroxides are rapidly decomposed by metal ions. For example, in the presence of copper salts in acetic acid, *n*-valeroyl peroxide is smoothly decomposed to 1-butene, valeric acid, and CO_2 (*71*) [Eqs. (63)–(65)]. The kinetics of the decomposition of valeroyl

$$(C_4H_9CO_2)_2 + \text{Cu(I)} \longrightarrow C_4H_9CO_2\cdot + \text{Cu(II)}(O_2CC_4H_9) \quad (63)$$

$$C_4H_9CO_2\cdot \longrightarrow C_4H_9\cdot + CO_2 \quad (64)$$

$$C_4H_9\cdot + \text{Cu(II)}(O_2CC_4H_9) \longrightarrow C_4H_8 + C_4H_9CO_2H + \text{Cu(I)} \quad (65)$$

peroxide and of other diacyl peroxides have been reported (*67,68,71,75–77*).

Synthetic use of simple aliphatic diacyl peroxides is severely handicapped by this facile catalyzed decomposition and by the extremely rapid oxidation of the alkyl radical by Cu(II). Thus, most of the peroxide is destroyed via an unproductive process. This problem is somewhat attenuated by the use of diaroyl peroxides. The aryloxy radicals formed do not decarboxylate as readily. Of the aromatic diacyl peroxides, benzoyl peroxide is commercially available and the most widely used. Unlike alkoxy radicals, the benzoyloxy radical formed from decomposition by Cu(I) attacks an olefin such as 1-butene by addition to the double bond as well as by hydrogen abstraction (*75*) [Eqs. (66)–(69)]. With aryl acyloxy radicals another point of consideration is necessary. If the rate of abstraction of

$$(C_6H_5CO_2)_2 + Cu(I) \longrightarrow C_6H_5CO_2\cdot + Cu(II)O_2CC_6H_5 \tag{66}$$

$$C_6H_5CO_2\cdot + \text{1-butene} \longrightarrow C_6H_5CO_2\text{-}C_4H_8\cdot \quad \text{or} \quad C_6H_5CO_2H + \text{(allyl radical)} \tag{67}$$

$$C_6H_5CO_2\text{-}C_4H_8\cdot \begin{cases} \xrightarrow{C_4H_8} C_6H_5CO_2C_8H_{16}\cdot \longrightarrow \text{Products} \\ \xrightarrow{C_4H_8} C_6H_5CO_2\text{-}C_4H_9 + \text{(allyl radical)} \\ \xrightarrow{Cu(II)} Cu(I) + H^+ + C_6H_5CO_2\text{-}C_4H_7 \end{cases} \tag{68}$$

$$\text{(allyl radical)} + Cu(II)X \begin{cases} \xrightarrow{85\%} CH_2{=}CH{-}CH(X){-}CH_3 + Cu(I) \\ \xrightarrow{15\%} CH_3{-}CH{=}CH{-}CH_2X + Cu(I) \end{cases} \tag{69}$$

the reactive hydrogen is of the same order of magnitude as the rate of loss of CO_2 from the acyloxy radical, a mixture of products will result not only from the reaction outlined above but as a result of secondary reactions of the aryl radical produced by loss of CO_2. In the example shown in Eq. (70), in which benzoyl peroxide was employed, products derived from both phenyl and benzoyloxy radicals were obtained (*13*).

$$CH_3(CH_2)_5CH{=}CH_2 + (C_6H_5CO_2)_2 \xrightarrow[C_6H_6]{Cu(I)} \underset{45\%}{CH_3(CH_2)_4CH_2CH{=}CH{-}C_6H_5} + CH_3(CH_2)_4CH(O_2CC_6H_5){-}CH{=}CH_2 + CH_3(CH_2)_4CH{=}CH{-}CH_2{-}O_2CC_6H_5 \tag{70}$$

The use of metal halides can cause further complications. Kochi and Subramanian have found that hypohalites are formed by reaction of peroxide and halide ion (*76*). The hypohalites produced in this manner scavenge the cuprous salt effectively, thus destroying the chain decomposition of the peroxide [Eqs. (71)–(73)]. **Because of the complications outlined above, diacyl peroxides are not peroxidic reagents of choice.**

$$(C_4H_9CO_2)_2 + Br^- \longrightarrow C_4H_9CO_2Br + C_4H_9CO_2^- \quad (71)$$

$$C_4H_9CO_2Br + Br^- \longrightarrow C_4H_9CO_2^- + Br_2 \quad (72)$$

$$2Cu(I) + Br_2 \rightleftharpoons 2Cu(II) + 2Br^- \quad (73)$$

3. Peresters

Peresters have been the most widely used peroxidic compounds for metal ion-catalyzed oxidations. Two peresters, *t*-butylperacetate and *t*-butylperbenzoate are commercially available from several sources, and both have been extensively employed. Two modes of reduction by a metal ion are possible [Eq. (74)].

$$t\text{-BuOOC}(=O)C_6H_5 + M^{+n} \begin{cases} \xrightarrow{(a)} t\text{-BuO}\cdot + M^{+n+1}(O_2CC_6H_5) \\ \xrightarrow{(b)} C_6H_5CO_2\cdot + M^{+n+1}(Ot\text{-Bu}) \end{cases} \quad (74)$$

Pathway (a), the reaction producing *t*-butoxy radicals, is energetically the more favorable reaction and the one observed (*72*). The simple *t*-butyl peresters are the most stable and the least prone to undergo rearrangement such as that shown in Eq. (75) (*5,6*). If primary or secondary groups

$$H_3C{-}C(CH_3)(R'){-}O{-}O{-}X \longrightarrow (H_3C)(R')C(OCH_3)(OX) \quad (75)$$

are used instead of a tertiary group, breakdown of the perester (or other peroxides) can occur by loss of hydrogen (*5,6*). Also, if tertiary alkyl groups other than *t*-butyl are employed, fragmentation can occur as described for hydroperoxides. For example, *t*-amylperbenzoate is decomposed smoothly to benzoic acid, acetone, and ethylene (*72*). Fragmentation of the *t*-amyloxy radical to acetone and ethylene occurs much faster than hydrogen abstraction [Eqs. (76)–(78)].

$$CH_3CH_2-\overset{\overset{CH_3}{|}}{\underset{\underset{CH_3}{|}}{C}}-OO\overset{\overset{O}{\|}}{C}C_6H_5 + Cu(I) \longrightarrow CH_3CH_2\overset{\overset{CH_3}{|}}{\underset{\underset{CH_3}{|}}{C}}-O\cdot + Cu(II)(O_2CC_6H_5) \quad (76)$$

$$CH_3CH_2\overset{\overset{CH_3}{|}}{\underset{\underset{CH_3}{|}}{C}}-O\cdot \longrightarrow CH_3CH_2\cdot + CH_3\overset{\overset{O}{\|}}{C}CH_3 \quad (77)$$

$$CH_3CH_2\cdot + Cu(II)(O_2CC_6H_5) \longrightarrow CH_2{=}CH_2 + C_6H_5CO_2H + Cu(I) \quad (78)$$

4. Dialkyl Peroxides

t-Butylperoxide is the most thermally stable peroxidic material commonly used. It too is commercially available. Primary and secondary peroxides are considerably less stable and decompose by nonchain-producing pathways, as described above for peresters. The same objections raised for tertiary groups other than *t*-butyl apply to dialkyl peroxides as well. Kochi (*72*) has reported the facile copper ion-catalyzed decomposition of higher dialkyl peroxides, thus adequately demonstrating the objections outlined above.

5. Peracids

Peracids are the last class of peroxidic materials that will be considered. Peracetic acid has been widely used commercially. Other peracids such as peroxypivalic (*78*), performic (*79*), peroxymonophthalic (*80*), trifluoroperacetic (*81*), and perbenzoic (*81*) have been described in the literature. **In general, care must be exercised in the handling of peracids, especially those of the lower aliphatic series. Traces of transition metal ions (Fe, Co, Ni, Mn, V) can cause very rapid decomposition of the pure, unstabilized acid (even as a 25% solution) with explosive violence. Thus reactions are carried out below 0° by the dropwise addition of the peracid solution to the reaction solution containing the compound to be oxidized and the catalyst. The maintenance of high dilution helps to minimize a run-away reaction.** Since peracids are excellent epoxidation reagents (*82*), substrates containing double bonds or those which have the capability of forming olefinic bonds during oxidation will often be converted to epoxidation products. For example, 6-methyl-2-piperidone is oxidized mainly to 6-methyl-3,4-dihydro-2-pyridone (**7**) (*83*). However, several secondary products are formed, including 6-methyl-5,6-dihydroxy-2-piperidone (**8**) and 6-methyl-5-oxo-2-piperidone (**9**) [Eq. (79)]. **The main advantage of the peracid route is the speed with which an oxidation can be accomplished.** For example,

$$\xrightarrow[0^\circ\ \mathrm{Mn(III)}]{CH_3CO_3H}$$

(**7**) + (**8**) + (**9**) (79)

2-piperidone is oxidized cleanly to glutarimide in high yield in 4 hr at -10°, whereas the same reaction using *t*-butylhydroperoxide at room temperature takes several days to reach completion.

The best catalysts for the oxidation of lactams are also among the best-known catalysts for decomposing peracids. The rate of oxidation of 2-piperidone decreases in the following order of metal ion catalysts: $Mn(II) \gg Co(II) \gg V(III)$. This same sequence of catalytic activity has been reported for the decomposition of peroxy radicals and peracids.

Brandon and Elliott (*41*) observed that the stability of peroxy radicals was dependent on the type of metal species in solution; this stability reportedly decreased in the following order: $Mn > Co > V$. Japanese workers found that peracrylic acid was decomposed by various metal ions with the following ease: $Mn > Co > Fe > Ni$ (*84*). The activation energy for the decomposition of peracetic acid was found to be in the order of 30–34 kcal/mole (*85,86*). *p*-Methylperoxybenzoic, peroxybenzoic, and *m*-chloroperoxybenzoic exhibited activation energies for the nonmetal ion-catalyzed decomposition of 16.6, 17.3, and 17.0 kcal/mole, respectively. The cobalt acetate-catalyzed decomposition of peracetic acid was reported by Koubek and Edwards (*87*) to proceed with an activation energy of 21.6 ± 1 kcal/mole. These authors (*87*) also isolated "cobaltic acetate" and postulated the structure of this compound as a dimer containing two cobalt ions with acetate ion serving as a bidentate ligand. The specific activity of manganese, cobalt, iron, and vanadium ions still cannot be categorized in any manner other than that they are all members of the first row of transition metals. However, the low energy of activation for the decomposition of peroxy acids with metal ions lends a ready explanation for the selectivity of this oxidative process and a rationale of why a slight increase in steric interference causes low yields or completely inhibits oxidation. **If the rate of oxidation is not comparable in rate to the decomposition of peroxyacid, no oxidation occurs.**

C. Experimental Conditions

1. Hydrocarbons

Useful synthetic oxidations of aliphatic and aromatic hydrocarbons have been accomplished mainly with alkyl hydroperoxides and dialkyl peroxides.

Unsymmetrical peroxides are most commonly produced during the oxidation of aliphatic and aromatic hydrocarbons with alkyl hydroperoxides. Cumene, for example, is smoothly oxidized with *t*-butylhydroperoxide and Co(II) to the mixed peroxide (*18*) [Eq. (80)]. Aliphatic

$$C_6H_5CH(CH_3)_2 + t\text{-BuOOH} \xrightarrow{\text{Co(II)}} C_6H_5C(CH_3)_2\text{—OO}t\text{-Bu} \quad (92\%) \qquad (80)$$

straight-chain hydrocarbons give the expected mixture of products resulting from a radical oxidation. Synthetically useful oxidations have been carried out, however, on the cyclic hydrocarbons. Evnin and Lam (*24*) have successfully oxidized cycloheptane with *t*-butylperoxide in the presence of Cu(II) and benzamide to provide *N*-cycloheptylbenzamide as the major

$$\text{cycloheptane} + (t\text{-BuO})_2 + C_6H_5CONH_2 \xrightarrow[C_6H_5Cl]{CuCl_2(phen)_2} \text{cycloheptyl-NHCOC}_6H_5 \ (68\%) + \text{cycloheptene} \qquad (81)$$

product [Eq. (81)]. Table II lists other examples of these types of reactions.

a. Oxidation with t-Butylhydroperoxide. **A mixture of 0.4 mole of the hydrocarbon in 0.84 mole of acetic acid containing 0.14 mole of *t*-butylhydroperoxide and 0.006 mole of manganous bromide is warmed at *50–100°***

TABLE II

Oxidation of Hydrocarbons

Reactant	Peroxide	Metal ion	Products	Yield, %	Reference
Adamantane	*t*-BuOO*t*-Bu	Cu (I) Cl/benzamide	1-$NHCOC_6H_5$-adamantane 39% + 15% 2°	54	*88*
$C_6H_5CH_3$	*t*-BuOOH *t*-BuOO*t*-Bu	Cu (I) Cl/phthalimide	N-benzylphthalimide ($N-CH_2-C_6H_5$)	75	*88,89*
	t-BuOO*t*-Bu	Cu (II) Cl_2/benzamide/HCl	$C_6H_5-CH_2-CH_2-C_6H_5$ and	60–80 75	*88*
			$C_6H_5-CH_2-NHCO-C_6H_5$	15	*88*
	t-BuOO*t*-Bu	Cu (I) Cl/saccharin		10	*88*

Continued

TABLE II—*Continued*

Reactant	Peroxide	Metal ion	Products	Yield, %	Reference
			SO_2, N—CH_2—, O	40	
			SO_2, N, OCH_2—		
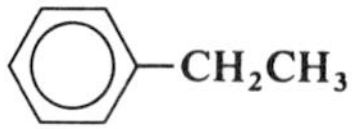 —CH_2CH_3	CH_3CO_3H	Mn (III)	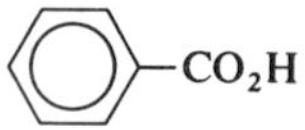 —CO_2H	60	*90*
	HOOH	Fe (II) SO_4	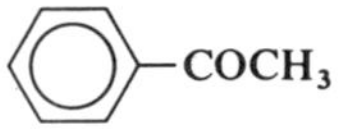 —$COCH_3$	—	*91*
	t-BuOO*t*-Bu	Cu (I)Cl/phthalimide	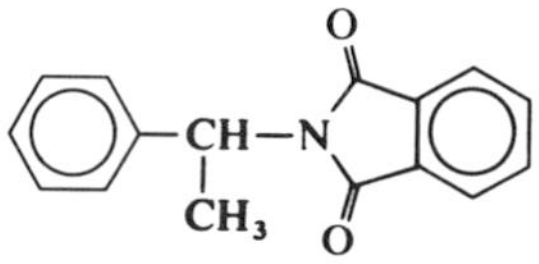 —CH—N, CH_3, O, O	42	*88*

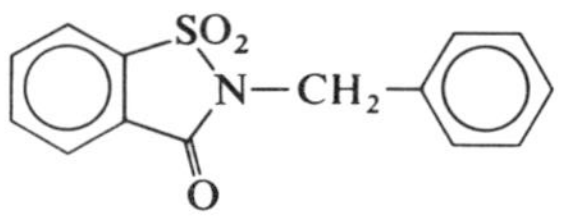

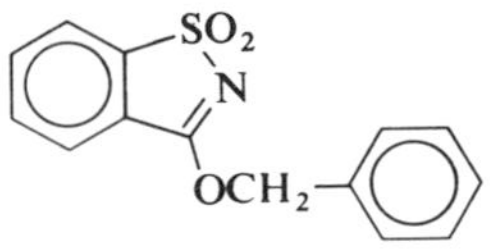

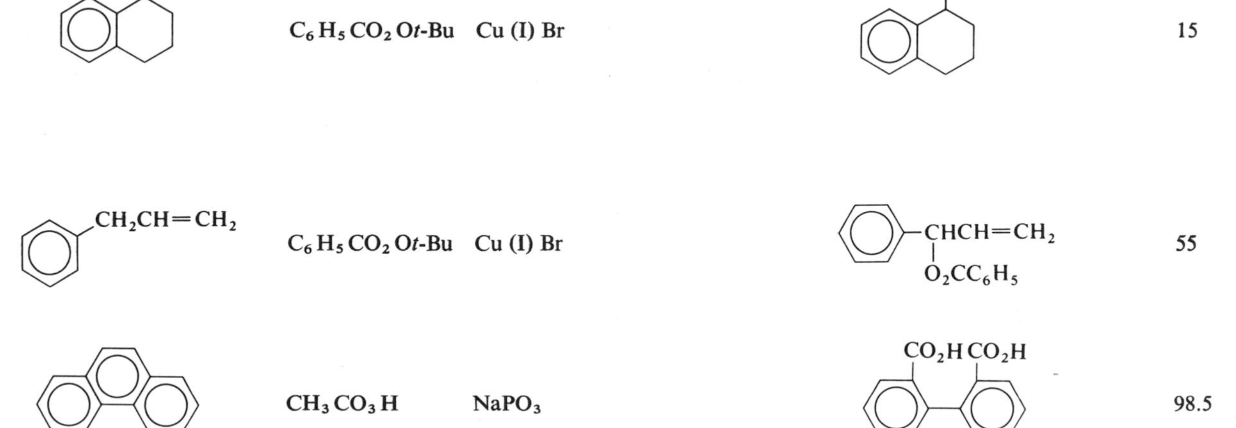

	$C_6H_5CO_2Ot\text{-}Bu$	Cu (I) Br		15	*92*
	$C_6H_5CO_2Ot\text{-}Bu$	Cu (I) Br		55	*17*
	CH_3CO_3H	$NaPO_3$		98.5	*93*

for 8–15 hr. After being poured into water, the organic phase is separated and worked up in an appropriate manner.

b. Oxidation with t-Butylperoxide in the Presence of Benzamide. **A mixture of 0.25 mole of benzamide and 0.5 mmole of cupric chloride–phenanthroline is heated in the reaction flask at 100° for 1 hr under vacuum (0.03 mm). Chlorobenzene (30 ml), 0.25 mole of hydrocarbon, and 50 mmoles of *t*-butylperoxide are added and the resulting mixture is heated at 100–120° for 24 hr. On cooling most of the unreacted benzamide precipitates.**

In these reactions the disappearance of *t*-butylperoxide can be conveniently followed by gas–liquid chromatography (GLC) using a 20 foot × 0.25-inch 20% silicone rubber column at 85°.

2. Olefins

Olefins have received the greatest amount of study. Peresters, alkyl hydroperoxides, and dialkyl peroxides have been used to synthetic advantage. As in the case of the hydrocarbons, the use of alkyl hydroperoxides can lead to a mixed peroxide as the major oxidation product. For example, cyclohexene is oxidized to a 90% yield of the mixed peroxide using *alpha*-cumylhydroperoxide (*18*) [Eq. (82)]. Cyclohexenylbenzoate is obtained in over 90% yield when the reaction is run using *t*-butylhydroperoxide in the presence of benzoic acid (*14*). *t*-Butylperacetate was used to oxidize *alpha*-pinene to a mixture of verbenylacetate and pinocarveol

$$\text{cyclohexene} + C_6H_5C(CH_3)_2OOH \xrightarrow{Cu(I)} \text{cyclohexenyl}{-}O_2C(CH_3)_2{-}C_6H_5 \quad (90\%) \qquad (82)$$

acetate in good yield (*94*) [Eq. (83)].

$$\alpha\text{-pinene} + t\text{-}BuOO_2CCH_3 \xrightarrow[HOAc\text{-}t\text{-}BuOH]{Cu(I)} \text{verbenyl-OAc} + \text{pinocarvyl-OAc} \qquad (83)$$

Amidation of olefins can also be accomplished with dialkyl peroxides. Evnin (*88*) recently reported that 4,4-dimethylpent-1-ene was oxidized with *t*-butylperoxide in the presence of phthalimide and Cu(II) to give a fair yield of *N*-(4,4-dimethylpent-2-en-lyl)phthalimide [Eq. (84)].

$$\text{olefin} + \text{phthalimide (NH)} + (t\text{-BuO})_2 \xrightarrow{\text{Cu(II)}} \text{N-allylphthalimide} \quad (84)$$

A number of reactions of olefins have been described in previous sections of this chapter (*13–19,44–47,65,66,75,76*); and other examples are listed in Table III.

a. Oxidation with t-Butylperacetate. **A solution of 0.5 mole of the olefin in a mixture of 40 ml of acetic acid and 40 ml of *t*-butyl alcohol containing 79.2 g of *t*-butylperacetate (75% in paraffin) and 1.0 g CuBr is heated at 60–80° for 12–16 h. The solvents are evaporated and the product isolated in an appropriate manner.**

b. Oxidation with t-Butylperoxide in the Presence of Phthalimide. **A mixture of 50 mmoles of olefin, 0.25 mole of phthalimide, 50 mmoles of *t*-butylperoxide, and 0.5 mmole of cupric chloride in 35 ml of chlorobenzene is heated at 80–100° for a prolonged time (2–5 days). The disappearance of *t*-butylperoxide is monitored by GLC. (See Section III.C.1b.). On completion of the reaction the mixture is cooled and unreacted phthalimide separates. The product amide is generally isolated by chromatography.**

3. Amides

Amides and lactams are readily oxidized with peracids and alkyl hydroperoxides to the corresponding imides. Peracid oxidation is generally faster, although undesired side reactions can occur (see Section III.B.5). Substitution on the amide nitrogen in the lactams appears to have little effect on the facility of the reaction, with *N*-methylpyrolidone being converted to *N*-methylsuccinimide about as easily as pyrrolidone itself is oxidized to succinimide (*25,83*). 2-Piperidone readily gives glutarimide on peracetic acid oxidation (*25,83*). Acetylated cyclic amines such as *N*-acetylpiperidine or *N*-acetylpyrrolidine were not oxidized with peracetic acid (*90*). The *N*-formyl derivative of piperidine was oxidized to only a small extent to *N*-formyl-2-piperidone (*25,83*). Acyclic *N*-alkylated amides are also readily oxidized, peracids again being the more suitable oxidant (*25,83*).

A mixture of 0.5 mole of the amide and 50 mg of manganous chloride in 300 ml of acetonitrile is cooled to −10°. Peracetic acid (1.5 moles) is added dropwise maintaining the temperature at −10°. The reaction terminates when the peroxide test is negative. The solution is filtered and the solvent is evaporated. The product is then isolated in an appropriate manner.

TABLE III

Oxidation of Olefins

Reactant	Peroxide	Metal ion	Products	Yield, %	Reference
$Me_3SiCH_2CH{=}CH_2$	$C_6H_5CO_2Ot\text{-}Bu$	Cu (I) Cl	$Me_3SiCH(OCOC_6H_5)CH{=}CH_2$	—	*95*
$Me_3SiOCH_2CH{=}CH_2$	$C_6H_5CO_2Ot\text{-}Bu$	Cu (I) Cl	$Me_3SiOCH(OCOC_6H_5)CH{=}CH_2$	15	*95*
	$C_6H_5C(CH_3)_2OOH$	Co (II) stearate	$—O—O—C(CH_3)_2—C_6H_5$	42	*96*
	$C_6H_5CO_2Ot\text{-}Bu$	Cu (I) Br	$—O_2CC_6H_5$	73	*15,17*
	$t\text{-}BuOOt\text{-}Bu$	Cu (I), Cu (II)/phthalimide	N, O, O	70	*89*

Substrate	Perester	Catalyst	Product	Yield	Ref.
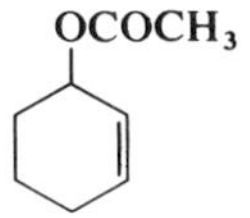 $OCOCH_3$	$C_6H_5CO_2Ot$-Bu	Cu (I) Br	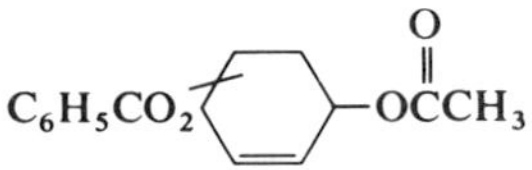 $C_6H_5CO_2$ … $OCCH_3$ (O)	29	*17*
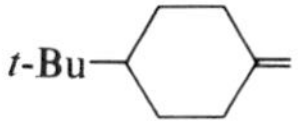 t-Bu–	$C_6H_5CO_2Ot$-Bu	Cu (I) Cl	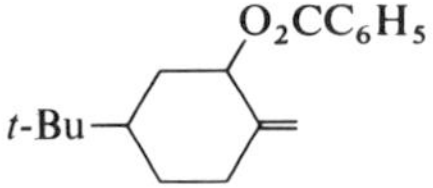 $O_2CC_6H_5$, t-Bu–	26	*49*
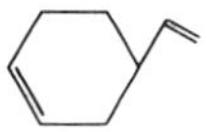	t-BuOOH	Co (II) naphthenate	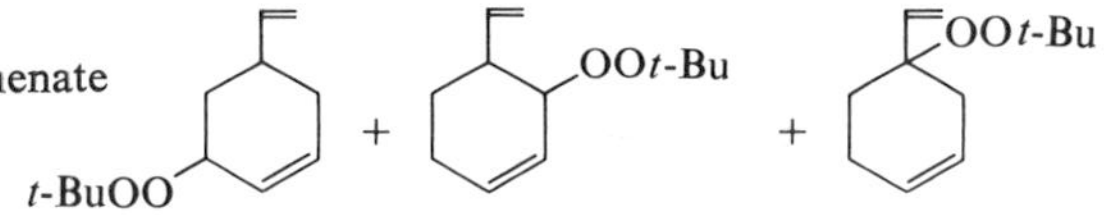 t-BuOO 75–85 + OOt-Bu 2–15 + OOt-Bu 1–3	51–52	*97*
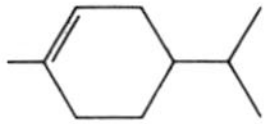 Opt. act.	t-$BuOO_2CCH_3$	Cu (II)	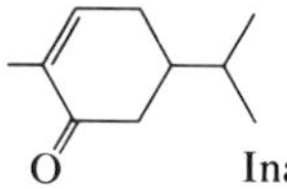 O Inactive	65.4	*47*

Continued

TABLE III—*Continued*

Reactant	Peroxide	Metal ion	Products	Yield, %	Reference
	$C_6H_5C(CH_3)_2OOH$	Co (II) stearate	$-OOC(CH_3)_2-C_6H_5$	15	*96*
	$C_6H_5CO_2Ot$-Bu	Cu (I) Br	Ot-Bu	25	*57,58,60*
	CH_3CO_2Ot-Bu	Cu (2-ethylhexoate)	O_2CCH_3 71 ± 3 + O_2CCH_3 29 ± 3	85	*44*

4. Sulfides and Sulfoxides

Sulfides are readily oxidized by alkyl hydroperoxides, peresters, and peracids. All three oxidants have distinct advantages. Alkyl hydroperoxides and peracids attack the sulfur atom producing sulfoxides and sulfones (26,90,98,99), whereas the peresters oxidize the position *alpha* to the sulfur atom giving *alpha*-thioesters (21,90,98,100–103) [Eq. (*85*)]. For example, dimethyl sulfide is oxidized rapidly to the sulfone with peracetic acid; no dimethyl sulfoxide is formed. Oxidation with *t*-butylhydroperoxide, however, gives predominantly the sulfoxide and very little sulfone (*25*).

Sulfoxides are readily converted to the sulfones with peracetic acid (*25*) [Eq. (85)]. Low yields of sulfones have been reported from some oxidations utilizing *t*-butylhydroperoxide with manganic acetylacetonate (*90*), but the use of cumene hydroperoxide with vanadium pentoxide reportedly gives good yields of the sulfone (*103a*).

$$CH_3SCH_3 \xrightarrow{CH_3CO_3H/Mn(III)} CH_3S(=O)_2CH_3$$

$$CH_3SCH_3 \xrightarrow{t\text{-}BuOOH/Mn(III)} CH_3S(=O)CH_3 + \text{trace } CH_3S(=O)_2CH_3 \qquad (85)$$

The oxidation of tetrahydrothiophene with *t*-butylperbenzoate provides a high yield of 2-benzoxyloxytetrahydrothiophene (*21*) [Eq. (86)].

$$\text{tetrahydrothiophene} + t\text{-}BuOO_2CC_6H_5 \longrightarrow \text{2-}(O_2CC_6H_5)\text{-tetrahydrothiophene} \;\; (69\%) \qquad (86)$$

Sulfones are not oxidized by either peracetic acid or *t*-butylhydroperoxide (*90*).

a. Oxidation with t-Butylhydroperoxide. **To a mixture of 0.1 mole of sufide and 50 mg of manganic acetylacetonate in 50 ml of acetonitrile was added dropwise 0.2 mole of *t*-butylhydroperoxide with ice cooling. The reaction is effectively completed at the end of the addition. The product sulfoxide is isolated by distillation.**

b. *Oxidation with t-Butylperbenzoate.* t-Butylperbenzoate (50 ml, 0.25 mole) is added to a mixture of the sulfide (0.5 mole), cuprous bromide (0.1 g, 0.35 mmole) and benzene maintained at 95–105° under nitrogen. After 6 hr, the reaction is cooled, diluted with ether (50 ml), and extıacted with 2 *N* sodium carbonate solution to remove benzoic acid. The etheral solution is washed with water, dried with anhydrous sodium sulfate, filtered, and concentrated.

5. Other Functional Groups

Peroxy *t*-butylbenzoate has been utilized for the oxidation of secondary alcohols to ketones in good yields (*104*), but the conversion of primary alcohols to aldehydes using this reagent is not very satisfactory (*92,101, 104*). Aldehydes are readily oxidized to acids by cumene hydroperoxide in the presence of osmium tetroxide or vanadium pentoxide (*7*). The use of other oxidants give anhydrides in poor to fair yields (*13,92*). Ketones give *alpha* peroxy ketones on oxidation with hydroperoxides, but the yields are generally poor except for *alpha* disubstituted ketones (*7,14,18*). *alpha* ketols, on the other hand, are readily cleaved to acids with hydroperoxides (*25*) as well as peracids (*90*).

As mentioned previously ethers are poor solvents for this type of oxidation reaction because of their reactivity. Reaction of ethers with peroxy *t*-butylbenzoate results in the replacement of an *alpha* hydrogen by a benzoate and/or *t*-butoxy group in generally good yields (*18,21,95,102, 105–112*). Some typical examples of this reaction are listed in Table IV.

Although nitriles (*16,17*) and esters (*17,92,107*) have been subjected to this oxidation procedure, the reactions appear to have little preparative value. However, this procedure has been utilized extensively to prepare amine oxides from tertiary amines in very good yields (*114–115*). In this reaction *t*-butylhydroperoxide is the oxidant of choice (*114*). Tertiary aromatic amines can give different types of products (*18,92,116*).

A mixture of 0.1 mole of the tertiary amine, 0.1 mole of *t*-butylhydroperoxide, and 0.05 g of vanadium oxyacetylacetonate in 30 ml of *t*-butyl alcohol is refluxed for 15 min. The mixture is cooled and the solvent is evaporated. The amine oxide is separated from the inorganic material by trituration with pentane or by extraction into water followed by evaporation.

ACKNOWLEDGMENTS

The author is pleased to acknowledge the help and encouragement of several colleagues during the preparation of this chapter. Drs. Anthony B. Evnin, David J. Trecker, Themistocles D. J. D'Silva, and Robert S. Neale offered unpublished results and suggestions for which the author is especially grateful.

APPENDIX

Cumene hydroperoxide

U.S. Peroxygen Division, Argus Chemical Corporation
850 Morton Ave.
Richmond, California 94804
Apogee Chemical, Inc.
De Carlo Avenue
Richmond, California
McKesson Chemical Co.
Charleston Branch
13th and Charles Avenue
P.O. Box 68
Dunbar, West Virginia
Lucidol Division, Pennwalt Corporation
1740 Military Road
Buffalo, New York 14240

t-Butylhydroperoxide

Lucidol Division, Pennwalt Corporation
The Norac Company, Inc.
405 South Motor Avenue
Azusa, California 91702
Aztec Chemicals
Division of Dart Industries, Inc.–Chemical Group
P.O. Box 756
Elyria, Ohio 44035
Apogee Chemical, Inc.
U.S. Peroxygen Division, Argus Chemical Corporation

t-Butylperacetate

Lucidol Division, Pennwalt Corporation
Aztec Chemicals

t-Butylperbenzoate

Lucidol Division, Pennwalt Corporation
The Norac Company, Inc.
Aztec Chemicals
Apogee Chemical, Inc.

t-Butylperoxide

Lucidol Division, Pennwalt Corporation
The Norac Company, Inc.
Aztec Chemical
Apogee Chemical, Inc.
U.S. Peroxygen, Division Argus Chemical Corp.

Peracetic acid (25% solution in ethyl acetate)

Union Carbide Corp.
127 Park Ave.
New York, New York

TABLE IV

Oxidation of Ethers

Reactant	Peroxide	Metal ion	Products	Yield, %	Reference
$(CH_3CH_2CH_2)_2O$	$C_6H_5CO_2Ot\text{-}Bu$	Cu (I) Br	$CH_3CH_2CH_2OCH(Ot\text{-}Bu)CH_2CH_3$ + $CH_3CH_2CH_2OCH(O_2C{-}C_6H_5){-}CH_2CH_3$	17 + 38.5	*105*
$((H_3C)_2CH)_2O$	$C_6H_5CO_2Ot\text{-}Bu$	Cu (I) Cl/butanol	$(H_3C)_2CH(O{-}n\text{-}Bu)_2$	31	*106,107*
	$C_6H_5CO_2Ot\text{-}Bu$	Cu (I) Cl/*n*-hexanol	$(H_3C)_2CH{-}(O{-}n\text{-}hexy)_2$	33	*107*
$(Ch_3CH_2CH_2CH_2)_2O$	$C_6H_5CO_2Ot\text{-}Bu$	Cu (I) Br	$CH_3(CH_2)_3O{-}CH(Ot\text{-}Bu)CH_2CH_2CH_3$	48	*109*
	$C_6H_5CO_2Ot\text{-}Bu$	Cu (I) Cl/*t*-BuOH	$CH_3(CH_2)_3OCH(Ot\text{-}Bu)CH_2CH_2CH_3$	48	*107*

	$C_6H_5CO_2Ot\text{-}Bu$	Cu (I) Br	$CH_3(CH_2)_3OCH(Ot\text{-}Bu)CH_2CH_2CH_3$ (27.5) + $CH_3(CH_2)_3OCH(O_2CC_6H_5)CH_2CH_2CH_3$ (40)	67.5	*105*
	$C_6H_5CO_2Ot\text{-}Bu$	Cu (I) Cl	$CH_3(CH_2)_3OCH(Ot\text{-}Bu)CH_2CH_2CH_3$ (15) + $CH_3(CH_2)_3OCH(O_2CC_6H_5)CH_2CH_2CH_3$ (50)	65	*102*
	$C_6H_5CO_2Ot\text{-}Bu$	Cu (I) Cl/*n*-hexanol	$CH_3(CH_2)_3OCH(OC_6H_{13})_2$	68–70	*106,107*
$CH_3(CH_2)_3OCH_2CH{=}CH_2$	$C_6H_5CO_2Ot\text{-}Bu$	Cu (I) Cl	$CH_3(CH_2)_3OCH(O_2CC_6H_5)CH{=}CH_2$	35	*102*
$C_6H_5OCH_2CH_3$	$C_6H_5CO_2Ot\text{-}Bu$	Cu (I) Cl	$C_6H_5OCH(O_2CC_6H_5){-}CH_3$	37	*102*
$C\ H_5CH_2OCH(CH_3)_2$	$C_6H_5CO_2Ot\text{-}Bu$	Cu (I) Cl	$C_6H_5CH(O_2CC_6H_5)OCH(CH_3)_2$	39	*112*

Continued

TABLE IV—*Continued*

Reactant	Peroxide	Metal ion	Products	Yield, %	Reference
$Me_3SiOCH_2CH_2CH_3$	CH_3CO_2Ot-Bu	Cu (I) Cl	$Me_3SiOCH(O_2CCH_3)CH_2CH_3$	—	*95*
	CH_3CO_2Ot-Bu	Cu (I) Br	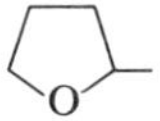 O*t*-Bu + trace 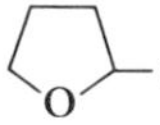 O_2CCH_3	41–45	*105,109*
	$C_6H_5CO_2Ot$-Bu	Cu (I) Br	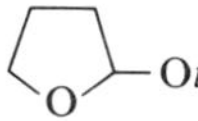 O*t*-Bu (45) 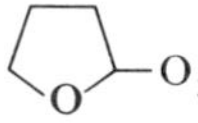 + $O_2CC_6H_5$ (26)	71	*105,109, 110*
	$C_6H_5CO_2Ot$-Bu	Cu (I) Cl/*t*-BuOH	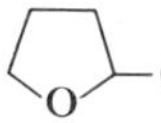 O*t*-Bu	57	*111*
	$C_6H_5CO_2Ot$-Bu	Cu (I) Cl/$C_6H_{13}OH$	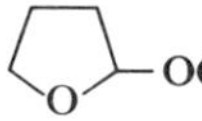 OC_6H_{13}	52	*111*
	$C_6H_5CO_2Ot$-Bu	Cu (I) Br	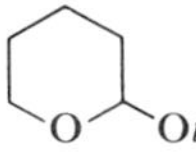 O*t*-Bu (33) 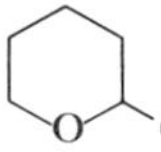 + $O_2CC_6H_5$ (24)	57	*105,109*
	$C_6H_5CO_2Ot$-Bu	Cu (I) Cl/*t*-BuOH	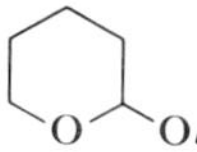 O*t*-Bu	38	*111*
	*t*-BuO*Ot*-Bu	Cu (II) $Cl_2(phen)_2$/ $C_6H_5CONH_2$	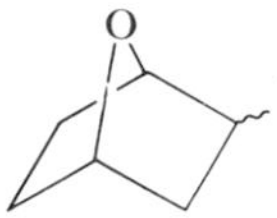 $NHCOC_6H_5$	20	*88*

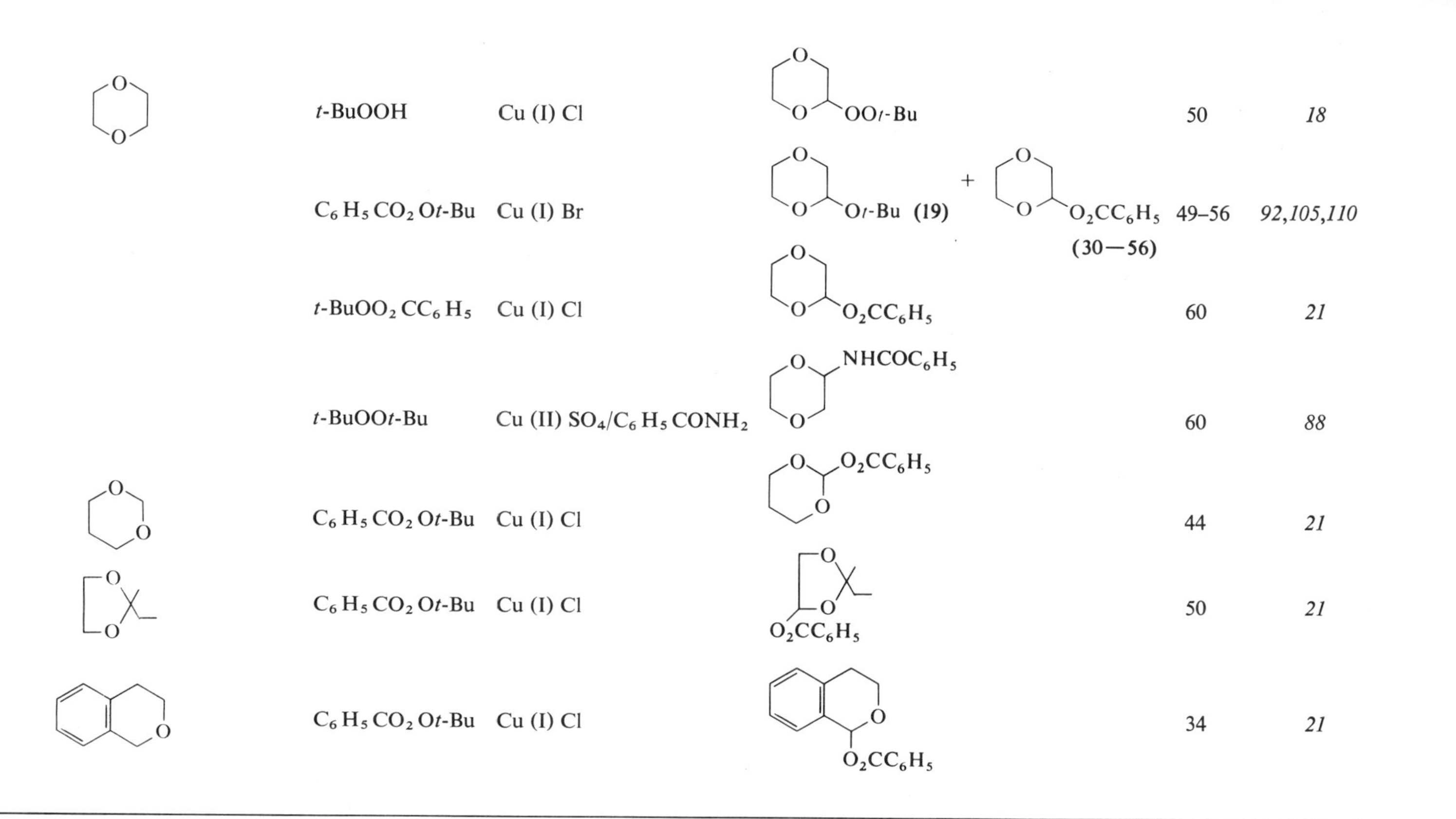

Substrate	Peroxide	Catalyst	Product	Yield	Ref.
	t-BuOOH	Cu (I) Cl	OOt-Bu	50	*18*
	$C_6H_5CO_2Ot$-Bu	Cu (I) Br	Ot-Bu (19) + $O_2CC_6H_5$ (30—56)	49–56	*92,105,110*
	t-BuOO$_2$CC$_6$H$_5$	Cu (I) Cl	$O_2CC_6H_5$	60	*21*
	t-BuOOt-Bu	Cu (II) $SO_4/C_6H_5CONH_2$	$NHCOC_6H_5$	60	*88*
	$C_6H_5CO_2Ot$-Bu	Cu (I) Cl	$O_2CC_6H_5$	44	*21*
	$C_6H_5CO_2Ot$-Bu	Cu (I) Cl	$O_2CC_6H_5$	50	*21*
	$C_6H_5CO_2Ot$-Bu	Cu (I) Cl	$O_2CC_6H_5$	34	*21*

REFERENCES

1. N. Uri, *Chem. Rev.*, **50,** 375 (1952).
2. W. A. Waters, *The Chemistry of Free Radicals*, Clarendon Press, Oxford, England, 1946, p. 247.
3. F. Haber and J. Weiss, *Naturwissenschaften*, **20,** 948 (1932).
4. J. Weiss, *Advan. Catal.*, **4,** 343 (1952).
5. A. G. Davies, *Organic Peroxides*, Butterworths, London, 1961, Chapters 12 and 13.
6. E. G. E. Hawkins, *Organic Peroxides, Their Formation and Reactions*, Van Nostrand, Princeton, N.J., 1961, Chapters 1, 3, and 7.
7. E. G. E. Hawkins, *J. Chem. Soc.*, 2169 (1950).
8. E. G. E. Hawkins and D. P. Young, *J. Chem. Soc.*, 2169, 2804 (1950).
9. E. G. E. Hawkins, *J. Chem. Soc.*, 3463 (1955).
10. M. S. Kharasch, A. Fono, W. Nudenberg, and B. Bischof, *J. Org. Chem.*, **17,** 207 (1952).
11. M. S. Kharasch, P. Pauson, and W. Nudenberg, *J. Org. Chem.*, **18,** 322 (1953).
12. M. S. Kharasch and W. Nudenberg, *J. Org. Chem.*, **19,** 1921 (1954).
13. M. S. Kharasch and A. Fono, *J. Org. Chem.*, **23,** 324 (1958).
14. M. S. Kharasch and A. Fono, *J. Org. Chem.*, **23,** 325 (1958).
15. M. S. Kharasch and G. Sosnovsky, *J. Amer. Chem. Soc.*, **80,** 756 (1958).
16. M. S. Kharasch and G. Sosnovsky, *Tetrahedron*, **3,** 105 (1958).
17. M. S. Kharasch, G. Sosnovsky, and N. C. Yang, *J. Amer. Chem. Soc.*, **81,** 5819 (1959).
18. M. S. Kharasch and A. Fono, *J. Org. Chem.*, **24,** 72 (1959).
19. M. S. Kharasch and A. Fono, *J. Org. Chem.*, **24,** 606 (1959).
20. G. Sosnovsky and S.-O. Lawesson, *Angew. Chem. Int. Ed.*, **3,** 269 (1964) and references cited therein.
21. C. Berglund and S.-O. Lawesson, *Ark. Kemi*, **20,** 225 (1964) and earlier papers.
22. J. K. Kochi and A. Bemis, *J. Amer. Chem. Soc.*, **90,** 4038 (1968).
23. J. K. Kochi, A. Bemis, and C. L. Jenkins, *J. Amer. Chem. Soc.*, **90,** 4616 (1968).
24. A. B. Evnin and A. V. Lam, *Chem. Commun.*, 1184 (1968) and unpublished results.
25. A. R. Doumaux, Jr., J. E. McKeon, and D. J. Trecker, *J. Amer. Chem. Soc.*, **91,** 3992 (1969).
26. W. A. Waters, 1964, *Mechanisms of Oxidation*, Wiley, New York.
27. R. Stewart, 1964, *Oxidation Mechanisms: Applications to Organic Chemistry*, Benjamin, New York.
28. K. B. Wilberg (ed.), *Oxidation in Organic Chemistry*, Part A, Academic Press, New York, 1965.
29. F. Basolo and R. G. Pearson, *Mechanisms of Inorganic Reactions, A Study o Metal Complexes in Solution*, Wiley, New York, 1967, Chapter 6.
30. H. Taube, *Advan. Inorg. Chem. Radiochem.*, **1,** 1 (1959).
31. I. M. Kolthoff, E. J. Meehan, M. S. Tsao, and Q. W. Choi, *J. Phys. Chem.*, **66,** 1233 (1962).
32. P. R. Carter and N. Davidson, *J. Phys. Chem.*, **56,** 877 (1952).
33. A. A. Clifford and W. A. Waters, *J. Chem. Soc.*, 2796 (1965).
34. F. Minisci and R. Galli, *Tetrahedron Lett.*, 357 (1963).
35. F. Minisci and R. Galli, *Tetrahedron Lett.*, 533 (1962).

36. H. E. De La Mare, J. K. Kochi, and F. F. Rust, *J. Amer. Chem. Soc.*, **85,** 1437 (1963).
37. D. K. Sebera and H. Taube, *J. Amer. Chem. Soc.*, **83,** 1785 (1961).
38. R. T. M. Fraser and H. Taube, *J. Amer. Chem. Soc.*, **81,** 5000 (1959).
39. J. Halpern and J. P. Candlin, *J. Amer. Chem. Soc.*, **85,** 2518 (1963).
40. M. Setaka, Y. Kirino, T. Ozawa, and T. Kwan, *J. Catal.*, **15,** 209 (1969) and references cited therein.
41. R. W. Brandon and C. S. Elliott, *Tetrahedron Lett.*, 4375 (1967).
42. M. Anbar and I. Pecht, *Int. Symp. Metal Organ.*, No. 9.
43. J. K. Kochi, *Rec. Chem. Progr.*, **27,** 207 (1966).
44. C. Walling and A. A. Zavitsas, *J. Amer. Chem. Soc.*, **85,** 2084 (1963).
45. D. B. Denney, D. Z. Denney and G. Feig, *Tetrahedron Lett.*, 19 (1959).
46. G. Sosnovsky and S.-O. Lawesson, *Angew. Chem. Int. Ed.*, **3,** 269 (1964), footnote 19.
47. D. Z. Denney, A. Appelbaum, and D. B. Denney, *J. Amer. Chem. Soc.*, **84,** 4969 (1962).
48. H. L. Goering and U. Mayer, *J. Amer. Chem. Soc.*, **86,** 3753 (1964).
49. B. Cross and G. H. Whitham, *J. Chem. Soc.*, 1650 (1961).
50. Reference *22*, footnote 21.
51. J. K. Kochi and D. M. Mog, *J. Amer. Chem. Soc.*, **87,** 522 (1965).
52. Kochi (*23*) assumes a very small isotope effect in the oxidative substitution reaction. Taking the ratio of k_e/k_s for both deuterium-substituted and -unsubstituted products and dividing gives a value of 2.8.
53. H. Gilman, R. G. Jones, and L. A. Woods, *J. Org. Chem.*, **17,** 1630 (1952).
54. G. M. Whitesides, W. F. Fischer, Jr., J. San Filippo, Jr., R. W. Bashe, and H. O. House, *J. Amer. Chem. Soc.*, **91,** 4871 (1969).
55. G. M. Whitesides, J. San Filippo, Jr., C. P. Casey, and E. J. Panek, *J. Amer. Chem. Soc.*, **89,** 5302 (1967).
56. E. J. Corey and G. H. Posner, *J. Amer. Chem. Soc.*, **89,** 3911 (1967); **90,** 5615 (1968).
57. P. R. Story, *J. Amer. Chem. Soc.*, **82,** 2085 (1960).
58. P. R. Story, *J. Org. Chem.*, **26,** 287 (1961).
59. P. R. Story, *J. Amer. Chem. Soc.*, **83,** 3347 (1961); *Tetrahedron Lett.*, No. 9, 401 (1962).
60. H. Tanida and T. Tsuji, *Chem. Ind. (London)*, 211 (1963).
61. H. Tanida and T. Tsuji, *J. Org. Chem.*, **29,** 849 (1964).
62. G. A. Olah, S. J. Kuhn, and S. H. Flood, *J. Amer. Chem. Soc.*, **84,** 1688 (1962).
63. G. A. Olah, N. A. Overchuk, and J. C. Lapierre, *J. Amer. Chem. Soc.*, **87,** 5785 (1965).
64. J. K. Kochi, *J. Amer. Chem. Soc.*, **83,** 3162 (1961).
65. J. K. Kochi, *J. Amer. Chem. Soc.*, **84,** 774 (1962).
66. J. K. Kochi, *J. Amer. Chem. Soc.*, **84,** 3271 (1962).
67. J. K. Kochi and H. E. Mains, *J. Org. Chem.*, **30,** 1862 (1965).
68. J. K. Kochi and R. V. Subramanian, *J. Amer. Chem. Soc.*, **87,** 4855 (1965), pp. 4859–4863.
69. J. M. Harvilchuck, D. A. Aikens, and R. C. Murray, Jr., *Inorg. Chem.*, **8,** 539 (1969).
70. B. W. Cook, R. G. J. Miller, and P. F. Todd, *J. Organometal. Chem.*, **19,** 421 (1969).

71. J. K. Kochi and A. Bemis, *Tetrahedron*, **24,** 5099 (1968).
72. J. K. Kochi, *Tetrahedron*, **18,** 483 (1962).
73. J. K. Kochi, *J. Amer. Chem. Soc.*, **85,** 1958 (1963).
74. J. K. Kochi, *J. Amer. Chem. Soc.*, **84,** 2785 (1962).
75. J. K. Kochi, *J. Amer. Chem. Soc.*, **84,** 1572 (1962).
76. J. K. Kochi and R. V. Subramanian, *J. Amer. Chem. Soc.*, **87,** 1508 (1965).
77. J. K. Kochi and R. D. Gilliom, *J. Amer. Chem. Soc.*, **86,** 5251 (1964).
78. E. Koubek and J. E. Welsch, *J. Org. Chem.*, **33,** 445 (1968).
79. G. Toennies and R. P. Homiller, *J. Amer. Chem. Soc.*, **64,** 3054 (1942).
80. H. Bohme, in *Organic Syntheses* (E. C. Horning, ed.), Coll. Vol. III, Wiley, New York, 1955, p. 619.
81. S. Lewis, in *Oxidation* (R. L. Augustine, ed.), Marcel Dekker, New York, 1969, pp. 213–258.
82. H. O. House, *Modern Synthetic Reactions*, Benjamin, New York, 1965.
83. A. R. Doumaux and D. J. Trecker, *J. Org. Chem.*, **35,** No. 7 (1970).
84. Y. Ohkatsu, T. Hara, T. Osa, and A. Misono, *Bull. Chem. Soc. Jap.*, **40,** 1893 (1967).
85. E. Koubek, M. L. Haggett, C. J. Battaglia, K. M. Ibne-Rasa, H. Y. Pyun, and J. O. Edwards, *J. Amer. Chem. Soc.*, **85,** 2263 (1963).
86. C. Schmidt and A. H. Sehon, *Can. J. Chem.*, **41,** 1819 (1963).
87. E. Koubek and J. O. Edwards, *J. Inorg. Nucl. Chem.*, **25,** 1401 (1963).
88. A. B. Evnin, unpublished results.
89. A. Fono, *Chem. Ind.* (*London*), 414 (1958).
90. A. R. Doumaux, Jr., unpublished results.
91. J. H. Merz and W. A. Waters, *J. Chem. Soc.*, 2427 (1949).
92. G. Sosnovsky and N. C. Yang, *J. Org. Chem.*, **25,** 899 (1960).
93. N. D. Rus'yanova and N. V. Malysheva, *Khim. Prom.*, **45,** 168 (1969).
94. T. D. J. D'Silva, unpublished results.
95. G. Sosnovsky and H. J. O'Neill, *Compt. Rend. Acad. Sci.*, **254,** 704 (1962).
96. W. Treibs and G. Pellmann, *Chem. Ber.*, **87,** 1201 (1954).
97. J. R. Shelton and J. N. Henderson, *J. Org. Chem.*, **26,** 2185 (1961).
98. Institut Francais du Petrok, des Carburants et Lubrifiants, French Patent 1,540,284 (1968).
99. Chem. Werke Huls A.G., Belgium Patent 679,240 (1966).
100. G. Sosnovsky, *Tetrahedron*, **18,** 15 (1962).
101. S.-O. Lawesson and C. Berglund, *Acta. Chem. Scand.*, **15,** 36 (1961).
102. S.-O. Lawesson, C. Berglund, and S. Grönwall, *Acta Chem. Scand.*, **15,** 249 (1961).
103. G. Sosnovsky, *J. Org. Chem.*, **26,** 281 (1961).
103a. L. Kuhnen, *Angew. Chem. Int.*, **5,** 893 (1966).
104. S.-O. Lawesson and C. Berglund, *Ark. Kemi*, **17,** 485 (1961).
105. G. Sosnovsky, *Tetrahedron*, **13,** 241 (1961).
106. S.-O. Lawesson and C. Berglund, *Angew. Chem.*, **73,** 65 (1961).
107. S.-O. Lawesson and C. Berglund, *Ark. Kemi*, **17,** 465 (1961).
108. S.-O. Lawesson and C. Berglund, *Tetrahedron Lett.*, No. 2, 4 (1960).
109. G. Sosnovsky, *J. Org. Chem.*, **25,** 874 (1960).
110. S.-O. Lawesson, C. Berglund, and S. Grönwall, *Acta Chem. Scand.*, **14,** 944 (1960).
111. S.-O. Lawesson and C. Berglund, *Ark. Kemi*, **17,** 475 (1961).
112. S.-O. Lawesson and C. Berglund, *Ark. Kemi*, **16,** 287 (1960).

113. J. E. McKeon and D. J. Trecker, unpublished results.
114. M. N. Sheng and J. G. Zajacek, *J. Org. Chem.*, **33,** 588 (1968).
115. L. Kuhnen, *Chem. Ber.*, **99,** 3384 (1966).
116. T. Caronna, G. P. Gardini, and F. Minisci, *Chem. Commun.*, 201 (1969).

AUTHOR INDEX

Numbers in parentheses are reference numbers and indicate that an author's work is referred to although his name is not cited in the text. Numbers in italics give the page on which the complete reference is listed.

A

Abeles, R. H., 26(81), 35(81), *60*
Abson, D., 32(114), *61*
Achenbach, H., 33(124), 34(124), *61*
Adams, W. R., 79(55), 98(111), 106(111), *110, 112*
Agami, C., 1(2), *58*
Aikens, D. A., 158(69), *183*
Akhtar, M., 24(74), 35(74), *60*
Albright, J. D., 5(30), 11(30), 17(30), 20(30), 24(30), 33(121), 34(121), 42(30, 121), 44(30), 45(30), 46(30), 47(30), 55(30), *59, 61*
Aldrich, T. B., 33(20), 50(120), *61*
Ali, Y., 31(107), 49(107), *61*
Allison, K. A., 116(24, 27), 117(27), 119(27), 120–125(24, 27), 130(27), 131(27), 133(27), *139*
Anbar, M., 145(42), *183*
Anderson, G. J., 1(6, 8), 13(6, 9), *59*
Antonaki, K., 32(111), 48(169, 170), 49(182), *61, 63*
Appel, R., 82(67), 84(80), 111
Appelbaum, A., 148(47), 171(47), 173(47), *183*
Arbuzov, Y. A., 69(17), *110*
Arens, J. F., 11(48), *60*
Arluck, R. M., 115(15, 16), *139*
Arndt, R., 33(123), 34(123), *61*
Arnold, S. J., 68(8, 10), *109*
Arvor, M. J., 49(182), *63*
Auerbach, J., 85(84), *111*

B

Badger, R. M., 68(9), *109*
Badoche, M., 74(39, 40), *110*
Bagli, J. E., 94(105, 106), 105(105, 106), *112*
Baizer, M. M., 1(10), 13(10), *59*
Baker, B. R., 28(98), 29(98), 30(106), *61*
Baker, D. A., 55(210), *64*
Balogh, V., 83(72), *111*
Ban, Y., 35(130), *62*
Barker, H. A., 24(77), *60*
Barnes, R. K., 123(39), 137(39), *140*
Barone, B. J., 115(18), *139*
Barrow, K. D., 53(199), *63*
Barton, D. H. R., 2(15), 13(15), 53(199), *59, 63*
Bashe, R. W., 153(54), *183*
Basolo, F., 143(29), 144, 146(29), *182*
Basselier, J.-J., 80(59, 59), 81(65), 93(103), *110, 111*
Battaglia, C. J., 165(85), *184*
Becker, H. D., 67(3), 94(3), *109*
Bell, R. A., 104(125), 105(125), *112*
Bemis, A., 142(22, 23), 148(22, 23), 149(22, 23), 150, 153(22, 23), 154, 160(71), 161(71), *182, 184*
Berglund, C., 142(21), 175(21, 101, 102), 176(101, 102, 104, 106, 107, 108, 110, 111, 112), 178(106, 107, 112), 179(102, 106, 107), 180(110, 111), 181(21, 110), *182, 184*
Bergman, W., 69(20), 93(20), *110*
Bernetti, R., 33(120), 50(120), *61*
Bernheim, F., 84(81), *111*
Berthelot, J., 81(62), *111*
Beyler, R. E., 23(72), *60*
Beynon, P. J., 24(76), 29(76), *60*
Bhatia, V. K., 51(192), *63*
Biemann, K., 33(124), 34(124), *61*
Bischof, B., 142(10), *182*

Bishop, C. E., 69(26), *110*
Bissing, D. E., 136(42), *140*
Blakley, R. L., 26(82), *60*
Bloch, J. C., 1(3), *58*
Bloomfield, J. J., 52(195), *63*
Blumberg, R., 113(4), 122(4), *138*
Bohlmann, F., 35(135), *62*
Bohme, H., 164(80), *184*
Boshart, G. L., 10(43), *59*
Bowen, E. J., 69(19), 70(27), 71(33, 34), 75(41, 42, 43), *110*
Bowers, A., 23(71), *60*
Bramer, R., 82(67), *111*
Brandon, R. W., 145(41), 165(41), *183*
Bredereck, K., 33(117), 50(117), *61*
Brenner, M., 101(114), *112*
Brill, W. F., 102(115), *112*, 115(17, 18, 19), 116(17), *139*
Brimacombe, J. S., 27(93), 48(176), *61*, *63*
Brodbeck, U., 32(116), 47(116), 48(116), *61*
Brook, A. G., 37(142, 143, 144), *62*
Brown, T. H., 53(202), *64*
Browne, R. J., 70(30), *110*
Brownson, C., 26(82), *60*
Brunken, J., 65(1), *109*
Bryan, J. G. H., 27(93), *61*
Büchi, G., 33(122), 34(122), 34(126, 127), 37(147), *61*, *62*
Buchwald, G., 95(107), *112*
Bukneeva, L. M., 114(9), 122(9), *138*
Burdon, M. G., 5(34, 35), 20(34, 35), 41(34, 35), 52(34), *59*
Burstain, I. G., 84(74), *111*
Buss, D. H., 28(98), 29(98), 30(106), *61*
Butenandt, A., 23(70), *60*

C

Calsoulacos, P., 48(178), 49(178), *63*
Candlin, J. P., 145(39), *183*
Capon, B., 17(63), *60*
Carnduff, J., 23(73), *60*
Carter, P. R., 144(32), *182*
Casey, C. P., 153(55), *183*
Cerny, M., 48(177), *63*
Chain, E. B., 53(199), *63*
Chambers, R. W., 68(11), 73(11), *109*
Charafi, M., 81(63), *111*
Chen, W. Y., 41(153), *62*
Chittenden, G. J. F., 48(168), 55(168), *63*
Choi, Q. W., 143(31), *182*
Church, J. M., 113(4), 122(4), *138*
Cilento, G., 13(53), *60*
Clark, B. C., Jr., 69(26), *110*
Clark-Lewis, J. W., 50(189), *63*
Claus, P., 53(203), *64*
Clifford, A. A., 144(33), *182*
Coffen, D. L., 33(122), 34(122, 127), *61*, *62*
Cohen, T., 2(13), 13(13), *59*
Collins, P. M., 24(76), 29(76), *60*
Comes, R. A., 35(133), *62*
Cone, N. J., 33(125), 34(125), *62*
Cook, A. F., 5(36), 27(88), 28(88), 29(88), 31(88), 32(115,), 41(36), 50(88),54(36), *59*, *61*
Cook, B. W., 158(70), *183*
Cooper, J. N., 16(62), *60*
Cope, A. C., 81(63), *111*
Corey, E. J., 35(134), *62*, 69(22), *110*, 153(56), *183*
Coronna, T., 176(116), *185*
Cramer, F., 11(47), *59*
Cree, G. M., 27(92), 29(103), 31(92), 48(103), 57(92), *61*
Cross, B. E., 50(190), *63* ,148(49), 173(49), *183*
Cruikshank, P. A., 10(43), *59*

D

Dalle, J. P., 107(132), 108(136, 137), *112*
D'Angeli, F., 40(151, 152), *62*
Dässler, H. G., 104(124), *112*
Davidson, N., 144(32), *182*
Davies, A. G., 141(5), 148(5), 163(5), *182*
Davis, C. C., 52(193), *63*
Davis, N. R., 37(143), *62*
de Belder, A. N., 33(119), 50(119), *61*
De La Mare, H. E., 144(36), *183*
Deletang, C., 93(103), *111*
deLong, D. C., 53(200), *63*
Delton, M., 32(112), *61*
de Mayo, P., 84(82), *111*
Demole, E., 81(66), 108(66), *111*
Denk, W., 103(119), 104(119), *112*

Denney, D. B., 147(45), 148(47), 171(45, 47), 173(47), *183*
Denney, D. Z., 147(45), 148(47), 171(45, 47), 173(47), *183*
Denny, R., 84(74), 102(116), *111*, *112*
Denson, D. D., 69(26), *110*
DeRoch, I. S., 116(26), 125(26), 131(26), 132(26), *139*
DiBello, C., 40(152), *62*
Dimant, E., 26(87), *61*
Djerassi, C., 23(71), 33(123), 34, 35(137), *60*, *61*, *62*
Dmitriev, B. A., 27(89), 50(89), *61*
Doganges, P. T., 24(76), 29(76), *60*
Doumaux, A. R., Jr., 83(71), *111*, 142(25), 164(83), 168(90), 171(25, 90), 172(90), 175(25, 90), 176(25, 90), *182*, *184*
Douraghi-Zadeh, K., 10(41), 12(41), *59*
Dowd, P., 36(140), *62*
Druckrey, E., 88(89), *111*
D'Silva, T. D. J., 170(94), *184*
Duff, J. M., 37(143) *62*
Dufraisse, C., 74(39, 40), 80(58, 59), 82(70), 88(90), 93(100, 101, 102), *110*, *111*
Dugan, J. J., 45(158), *62*
Dunlap, D. E., 77(48), 78(48), *110*
Durham, L. J., 49(185), *63*
Dyer, J. R., 28(97), *61*

E

Ecary, S., 82(70), *111*
Edwards, J. Q., 165(85, 87), *184*
Eggart, E., 50(188), *63*
Eggert, H., 103(119), 104(119), *112*
Elliott, C. S., 145(41), 165(41), *183*
Elliot, R. D., 37(146), *62*
Emoto, S., 48(175), *63*
Enggist, P., 81(66), 108(66), *111*
Engle, R. R., 23(71), *60*
Epstein, W. W., 1(5), 16(60), 21(60), 47(60), *59*, *60*
Eschenmoser, A., 22(68), *60*
Etienne, A., 80(59), *111*
Evnin, A. B., 142(24), 154(24), 157(24), 166, 167(88), 168(88), 170(88), 180 (88), 181(88), *182*, *184*

F

Farine, J.-C., 69(26), *110*
Fatiadi, A. J., 49(184), *63*
Feig, G., 147(45), 171(45), *183*
Feldkamp, J., 48(167), *63*
Felsenstein, A., 26(87), *61*
Fenical, W., 66(2), 99(2), 106(2), *109*
Fenselau, A. H., 5(33), 11(44), 13(56), 20(33), 41(33, 154), 53(33), *59*, *60*, *62*
Field, T. D., 32(114), *61*
Filira, F., 40(151, 152), *62*
Finlayson, N., 68(8), *109*
Finnegan, R. A., 115(12), *138*
Fischer, W. F., Jr., 153(54), *183*
Fisher, G. S., *112*
Fletcher, H. G., 27(94), 32(113), 47(164), 49(113, 181), 50(164), *61*, *63*
Flood, S. H., 157(62), *183*
Floyd, M. B., 86(87), 87(87), 89(93), 91(93), 108(93), *111*
Foner, S. N., 69(21), *110*
Fono, A., 142(10, 11, 12, 18, 19), 166(18), 167(89), 170(18), 171(18, 19), 176(18), 181(18), *182*, 184
Foote, C. S., 68(6, 7), 69(16), 70(6), 76(16), 84(74), 95(109), 99(112), 100(112), 101(114), 102(16, 115, 116, 117), 103(16), *109*, *111*, *112*
Forbes, E. J., 76(45), 80(60, 61), *110*, *111*
Foster, G., 116(27), 117(27), 119(27), 120–123(27), 125(27), 130(27), 131 (27), 133(27), *139*
Fox, J. E., 105(126), *112*
Franck, R. W., 85(84), *111*
Fraser, R. T. M., 145(36), *183*
Frey, P. A., 26(81), 35(81), *60*
Fried, J., 7(37), 21(37), *59*
Fripiat, J., 116(33b), 120(33b), 122(33b), 125(33b), 130, *139*
Fukami, H., 28(95, 96), *61*
Furutachi, N., 107(131), *112*

G

Gabriel, O., 48(179), 49(179), *63*
Gabriel, T., 41(153), *63*
Gadsby, G., 47(163), 50(163), *63*

Galli, R., 144(34, 35), *182*
Gardini, G. P., 176(116), *185*
Gardner, B. J., 2(15), 13(15), *59*
Gastrok, W. H., 38(149), 150(149), *62*
Gerard, M., 93(102), *111*
Ghambeer, R. K., 26(82), 35(82), *60*
Gilliom, R. D., 161(77), 163(77), *184*
Gilman, H., 153(53), *183*
Giormani, V., 40(151), *62*
Glinski, R. P., 55(208), *64*
Godman, J. L., 50(186), *63*
Goering, H. L., 148(48), *183*
Goldman, L., 5(30), 11(30), 17(30), 20(30), 24(30), 33(121), 34(121), 42(30, 131), 44(30), 45(30), 46(30), 47(30), 55(30), *59, 61*
Gollnick, K., 69(13, 14, 15), 71(13), 78(49), 81(13), 93(13, 14), 95(107, 109, 110), 100(13, 49), 102(13, 14, 15), 103(110), 104(14, 110), 105(129), 106(14), 107(14), *109, 110, 112*
Goodman, L., 48(172), *63*
Gould, E. S., 116(31, 32), 120(31, 32), 121(31, 32), 124–126(31, 32), 130(31, 32), 131(32), 133(32), 134(31), 135, 136, 137, *139*
Greene, F. D., 36(141), *62*
Greenspax, G., 47(163), 50(163), *63*
Grevels, F. W., 108(135), *112*
Griffiths, J., 76(45), 80(60, 71), *110, 111*
Grönwall, S., 175(102), 176(102, 110), 179(102), 180(110), 181(110), *184*
Gubler, B., 37(147), *62*
Guggisberg, A., 45(157), *62*
Gupta, S. K., 4(29), 11(29, 45), 18(64b), 47(162), *59, 60*

H

Haber, F., 141(3), *182*
Haggett, M. L., 165(85), *184*
Halpern, J., 145(39), *183*
Hamamura, E. H., 41(154), *62*
Hamsker, J., 23(72), *60*
Hara, T., 165(84), *184*
Harding, M. J. C., 88(92), *111*
Harmon, R. E., 4(29), 11(29), 18(64a, 64b), *59, 60*
Harvilchuck, J. M., 158(69), *183*
Hata, K., 44(156), *62*
Hawkins, E. G. E., 116(21), 141(6), 142(7, 8, 9), 148(6), 163(6), 176(7), *139, 182*
Hayashi, Y., 53(201), *63*
Hyeon, S. B., 109(138, 139), *112*
Heaton, L., 116(20), *139*
Henderson, J. N., 172(97), *184*
Henneberg, D., 38(148), *62*
Hesse, M., 45(157, 158), *62*
Hiatt, R., 116(31), 120(31), 121(31), 124–126(31), 130(31), 134(31, 41), 135, 136, 137, *139, 140*
Higgins, R., 102(117), *112*
Hirai, S., 35(128), 45(128), *62*
Hirano, S., 5(31), 24(31), 29(31), 44(31), 54(31, 206), 55(31), 56(31), *59, 64*
Hlavka, J. J., 10(42), *59*
Ho, P. T., 35(138), *62*
Hochrainer, A., 53(204), *64*
Hochstetler, A. R., 103(122), *112*
Hoffe, P., 35(131), *62*
Hoffman, A. K., 13(51), *60*
Hogenkamp, H. P. C., 24(77), 26(82), 35(82), 48(174), 49(174), *60, 63*
Hollins, R. A., 68(11, 12), 73(11, 12), *109*
Homiller, R. P., 164(79), *184*
Horecker, B. L., 35(131), *62*
Hori, Z., 45(159), *62*
Horton, D., 26(84, 85), 31(110), 32(110), 37(110), 44(110), 47(110), 48(110, 165), 50(110, 186), *61, 63*
Houghton, B., 51(191), *63*
Houpillard, J., 93(101), *111*
House, H. O., 153(54), 164(82), *183, 184*
Howard, J. A., 70(28), *110*
Howarth, G. B., 26(83), 50(83), *60*
Howgate, P., 4(25), *59*
Huber, J. E., 106(130), *112*
Huckstep, L. L., 33(125), 34(125), 53(200), *62, 63*
Hudson, R. L., 69(21), *110*
Huffman, J. W., 37(145), *62*
Huges, J. G., 26(85), *61*
Hughes, N. A., 48(166), *63*
Hunsberger, I. M., 1(7), 2(7), 13(7), *59*
Hunt, P. F., 24(74), 35(74), *60*

Husain, A., 27(93), *61*
Hussain, A., 48(176), *63*

I

Iacona, R. N., 1(12), *59*
Ibne-Rasa, K. M., 165(85), *184*
Ichikawa, H., 109(138), *112*
Ifzal, S. M., 46(160), 47(160), *62*
Indicator, N., 116–120(22), 122(22), 127 (22), 135(22), *139*
Ingold, K. U., 70(28), *110*
Inhoffen, H. H., 47(161), 53(161), *63*
Ireland, J. R. S., 52(195), *63*
Ireland, R. E., 104(125), 105(125), *112*
Irwin, K. C., 116(31), 120(31), 121(31), 124–126(31), 130(31), 134(31), 135, 136, 137, *139*
Ishadate, M., 31(108), *61*
Isoe, S., 109(138, 139), *112*
Iwashige, T., 26(86), *61*

J

Jager, P., 47(161), 53(161), *63*
James, K., 29(102), 48(102), *61*
Jautelat, J., 35(134), *62*
Jenkins, C. L., 142(23), 148(23), 149(23), 150, 153(23), *182*
Jewell, J. S., 31(110), 32(110), 37(110), 44(110), 47(110), 48(110, 165), 50(110), *61*, *63*
Johnson, C. R., 12(50), 13(55), 17(55), 20(55), 42(55), 43(55), *60*
Johnson, P., 116(27), 117(27), 119(27), 120–123(27), 125(27), 130(27), 131(27), 133(27), *139*
Jones, A. S., 4(25, 27, 29), *59*
Jones, D. N., 1(11), *59*
Jones, G. H., 25(78a, 78b, 79, 80), 26(78a, 78b), *60*
Jones, J. B., 9(38), 23(38), *59*
Jones, J. K. N., 26(83), 50(83), *60*
Jones, P. J., 37(143, 144), *62*
Jones, R. G., 153(53), *183*
Jones, W. J., 1(6, 8), 13(6, 8), *59*
Joseph, K. T., 50(187), *63*

K

Kabota, T., 53(198), *63*
Kalvoda, L., 48(177), *63*
Kaman, A. J., 114(11), 122(11), *138*
Kampe, W., 2(16), 13(16), *59*
Kampf, A., 26(87), *61*
Kaplan, M. L., 69(23, 24), *110*
Karliner, J., 35(137), *62*
Kasha, M., 70(29, 31), *110*
Kashimura, N., 5(31), 24(31), 29(31), 44(31), 54(31, 206), 55(31), 56(31), *59*, *64*
Kato, S., 116(29), 117(29), 118(29), 120–127(29), 131(29), 132(29), 135 (29), 136(29), *139*
Katz, T. J., 82(72), *111*
Kautsky, H., 68(4), *109*
Kawana, M., 48(175), *63*
Kawata, K., 35(128), 45(128), *62*
Kayima, T., 37(145), *62*
Kearns, D. R., 66(2), 68(11, 12), 73(11, 12), 99(2), 106(2), *109*
Keehn, P. M., 94(104), *112*
Kenny, R. L., *112*
Khan, A. U., 68(11, 12), 70(29, 31), 73 (11, 12), *109*, *110*
Kharasch, M. S., 142(10–19), 162(13), 163(17), 166(18), 170(4, 18), 171(13–19), 172(15, 17), 173(17), 176(13, 14, 16, 17), *182*
Khorana, H. G., 3(21), 10(40), 11(40), 12(40), *59*
Kiely, D. E., 49(181), *63*
Kinkel, K. G., 71(35), *110*
Kirino, Y., 145(40), *183*
Kirkwood, S., 27(90), *61*
Kitzing, R., 83(73), *111*
Klein, E., 38(148), 40(150), *62*, 103(120), 104(123), 105(123, 128), *112*
Koch, E., 82(68), *111*
Koch, H. J., 32(114), *61*
Kochetkov, N. K., 27(89), 50(89), *61*
Kochi, J. K., 142(22, 23), 144(36), 147(43), 148(22, 23, 43), 149(22, 23, 50, 51), 150, 153(22, 23, 52), 154, 157(64, 65), 158(65, 66, 67, 68), 160(67, 71–74),

161(51, 67, 68, 71, 75), 162(75), 164(72), 171(75), 172(66), *182*, *183*, *184*
Kocsis, K., 33(122), 34(122, 127), *61*, *62*
Koh, H. S., 28(95, 96), *61*
Kollar, J., 116(23), 120(23), 121(23), 123(23), 124(23), 125(23), 132(23), *139*
Kolthoff, I. M., 143(31), *182*
Kondo, K., 53(205), *64*
Kopecky, K. R., 76(46), *110*
Kornblum, N., 1(6, 8), 13(6, 8), *59*
Kost, A. A., 27(89), 50(89), *61*
Koubek, E., 164(78), 165(85, 87), *184*
Krauch, C. H., 67(3), 91(97), 94(3), *109*, *111*
Krishnamurty, H. C., 51(192), *63*
Krishna Rao, C. S., 50(187), *63*
Kuhn, S. J., 157(62), *183*
Kuhnen, L., 116(28), 117(28, 34, 37), 121(28), 122(28, 37), 132(28), *139*, *140*, 176(115), *185*
Kulsa, P., 134(126), *62*
Kuo, C. H., 21(67), *60*
Kurihara, N., 31(109), *61*
Kurihara, T., 45(159), *62*
Kurtz, D. W., 87(88), *111*
Kurzer, F., 10(41), 12(41), *59*
Kutsumura, S., 109(138), *112*
Kuzuhara, H., 32(113), 47(164), 49(113), 50(164), *61*, *63*
Kwan, T., 145(40), *183*

L

Ladd, J. N., 24(77), *60*
Lan, A. V., 142(24), 154(24), 157(24), *166*, *182*
Lane, D. C., 26(83), 50(83), *60*
Lapierre, J. C., 157(63), *183*
Larson, H. O., 1(6), 13(6), *59*
Lawesson, S.-O., 142(20, 21), 147(46), 171(46), 175(21, 101, 102), 176(101, 102, 104, 106, 107, 108, 110, 111, 112), 178(106, 107, 112), 179(102, 106, 107), 180(110), 181(21, 110, 111), *182*, *183*, *184*
Lawrence, R. J., 73(36), *110*
Lawrence, R. V., 100(113), *112*
Leclerq, F., 48(170), 49(182), *63*
Lee, W. W., 48(172), *63*
Leeming, M. R. C., 47(163), 50(163), *63*
Leitich, J., 2(17), *59*
Lerch, U., 5(33), 20(33), 41(33), 53(33), *59*
LeRoux, J.-P., 81(65), *111*
Levand, O., 1(6), 13(6), *59*
Lewis, S., 164(81), *184*
Liberles, A., 84(77), 85(77), *111*
Lillien, I., 12(49), *60*
Lin, J. W-P., 99(112), 100(112), *112*
Lin, K., 35(134), *62*
Lindberg, B., 33(119), 48(173), 50(119), *61*, *63*
Lindler, H., 9(39), *59*
List, F., 116(28), 117(28), 121(28), 122(28), 132(28), *139*
Linstead, R. P., 113(5), 122(5), *138*
Liss, T. A., 81(62), *111*
Lively, D. H., 53(200), *63*
Livingston, R., 69(18), 75(44), *110*
Lloyd, R. A., 70(27), 71(34), *110*

M

Mackie, D. W., 27(92), 31(92), 57(92), *61*
Maddox, M. L., 14(57), *60*
Madhav, R., 51(192), *63*
Mahlhop, R., 47(161), 53(161), *63*
Mains, H. E., 158(67), 160(67), 161(67), *183*
Malysheva, N. V., 169(93), *184*
Mander, L. N., 104(125), 105(125), *112*
Mani, J.-C., 107(132), 108(136, 137), *112*
March, M. M., 53(200), *63*
Marchand, A. P., 52(195), *63*
Machu, 114(6), 122(6), *138*
Marino, J. P., 20(66), *60*
Markham, R., 3(21), *59*
Marshall, J. A., 103(122), *112*
Martel, J., 84(75), 88(90), 91(75), *111*
Martin, D., 1(1), *58*
Mashio, F., 116(29), 117(29), 118(29), 120–127(29), 131(29), 132(29), 135(29), 136(29), *139*
Matsui, M., 31(108), *61*
Matsutani, S., 53(198), *63*
Matsuura, T., 90(94, 95), 96, 107(96), 108(134), *111*, *112*

Matuszak, C. A., 136(42), *140*
Mawanka, J., 116(33c), 122(33c), 125(33c), 130, *139*
Mayer, U., 148(48), *183*
Mayo, F. R., 115(15, 16), 134(41), *139*, *140*
McCant, D., 12(50), *60*
McCasland, G. E., 49(185), *63*
McDonald, E. J., 29(101), 48(101), *61*
McEwen, W. E., 136(42), *140*
McGonigal, W. E., 28(97), *61*
McKeown, E., 71(32), *110*, 142(25), 171(25), 175(25), 176(25, 113), *182*, *185*
McLean, M. J., 69(20), 93(20), *110*
Meehan, E. J., 143(31), *182*
Meinwald, J., 35(136), *62*
Mendelson, W. L., 73 (37,38), 105(127), 107(133), *110*, *112*
Mengler, C. D., 47(161), 53(161), *63*
Menguy, P., 116(26), 125(26), 131(26), 132(26), *139*
Mertens, H. J., 71(35), *110*
Merz, J. H., 168(91), *184*
Miethe, R., 35(135), *62*
Mikol, G. J., 13(54), 20(54), 42(54), *60*
Milas, N. A., 113(1a, 1b), 114(1a, 1b), *138*
Mill, T., 134(41), *140*
Miller, A. H., 86(86), *111*
Miller, R. G. J., 158(70), *183*
Minisci, F., 144(34, 35), 176(116), *182*, *185*
Misono, A., 165(84), *184*
Mitchell, R. E. J., 32(114), *61*
Moffatt, J. G., 2(18, 20), 3(20, 22), 5(33–36), 8(20), 11(20, 44), 12(20), 13(56), 14(57), 15(59), 19(65), 20(33–35), 21(65), 22(22), 23(65), 24(64), 25(20, 22, 78a, 78b, 79, 80), 26(78a, 78b), 27(88), 28(88), 29(88, 100), 31(88, 115, 116), 33(20, 65), 34(65), 37(65), 38(59), 39(59), 41(33–36, 54, 59), 47(65, 116), 48(116), 50, 52(24), 53(33), 54(36), 55(115), *59–60*
Mog, D. M., 149(51), 161(51), *183*
Monagle, J. J., 1(9), 13(9), *59*
Montgomery, J. A., 37(146), *62*
Moore, H. W., 53(196), *63*
Moore, R. N., 73(36), *110*
Morgan, J. E., 84(81), *111*
Morita, Z., 53(205), *64*
Moss, G. P., 4(24), *59*
Mousseron-Canet, M., 107(132), 108(136, 137), *112*
Mugden, M., 113(3), 122(3), *138*
Müller, W., 80(57), *111*
Murray, R. C., Jr., 158(69), *183*
Murray, R. W., 69(23, 24), *110*
Myers, P. L., 50(190), *63*

N

Nace, H. R., 1(9, 12), 13(9), *59*
Nader, F., 53(197), *63*
Nagai, M., 35(130), *62*
Nagarajan, K., 45(157), *62*
Nagarajan, R., 53(200), *63*
Nagata, W., 35(128), 45(128), *62*
Nakadaira, Y., 107(131), *112*
Nakadate, M., 26(84), *61*
Nakajima, M., 28(95, 96), 31(109), *61*
Nakanishi, K., 107(131), *112*
Nauman, M. O., 49(185), *63*
Neumüller, O. A., 78(49), 100(49), *110*
Neuss, N., 33(125), 34(125), 53(200), *62*, *63*
Newman, M. S., 52(193), *63*
Nickon, A., 73(37, 38), 94(105, 106), 105(105, 106, 107), 107(133), *110*, *112*
Niclas, H. J., 1(1), *58*
Nicolaides, N., 22(69), *60*
Ninomiya, I., 45(159), *62*
Novák, J. J. K., 28(99), *61*
Nozaki, H., 53(205), *64*
Nsumba, P., 116(33a), 122(33a), 125(33a), 130, *139*
Nudenberg, W., 142(10, 13, 14), 162(13), 170(14), 171(13, 14), 176(13, 14), *182*
Nussbaum, A. L., 41(153), *62*

O

Oda, R., 53(201), *63*
Ogryzlo, E. A., 68(8, 10), 70(30), *109*, *110*
Ohkatsu, Y., 165(84), *184*

Ohloff, G., 79(56), 95(107, 109), 103(118), *110, 112*
Ohnsorge, U. F. W., 53(199), *63*
Ohrui, H., 48(175), *63*
Oishi, T., 35(130), *62*
Okada, M., 31(108), *61*
Okumura, T., 35(128), 45(128), *62*
Olah, G. A., 157(62, 63), *183*
Olofson, R. A., 11(46), 20(66), *59, 60*
O'Neill, H. J., 172(95), 176(95), 180(95), *184*
Onodera, K., 5(31), 24(31), 29(31), 44(31), 54(31, 206), 55(31), 56(31), *59, 64*
Osa, T., 165(84), *184*
Overchuk, N. A., 157(63), *183*
Overend, W. G., 24(76), 29(76), *60*
Owen, L. N., 113(5), 122(5), *138*
Ozawa, T., 145(40), *183*

P

Pacak, J., 48(177), *63*
Panek, E. J., 153(55), *183*
Paquette, L. A., 36(139), *62*
Parikh, J. R., 5(32), 21(32), 44(32), 56(32), 57(32), *59*
Parker, A. J., 1(4), *58*
Parvez, M. A., 24(74), 35(74), *60*
Pascual, C., 50(188), *63*
Pauson, P., 142(11), *182*
Payne, G. B., 114(8a), 122(8a), *138*
Payne, J. Q., 115(13), *139*
Pearson, R. G., 143(29), 144, 146(29), *182*
Pecht, I., 145(42), *183*
Peedle, G. J. D., 37(144), *62*
Pelletier, S. W., 35(132), *62*
Pellmann, G., 172(96), *184*
Perkins, M. J., 17(63), *60*
Perlin, A. S., 27(92), 29(103), 31(92), 48(103), 57(92), *61*
Pettit, G. R., 47(162), 53(202), *63, 64*
Pfennig, H., 80(57), *110*
Pfitzner, K. E., 2(18, 20), 3(20, 22), 8(20), 11(20), 12(20), 19(65), 20(66), 21(65), 22(22), 23(65), 24(65), 25(20, 22), 33(20, 65), 34(65), 35(65), 37(65), 47(65), *59, 60*
Phillips, A., 35(138), *62*
Phillips, W. G., 13(55), 17(55), 20(55), 42(55), 43(55), *60*
Pierce, J. B., 37(142), *62*
Pobiner, H., 136(43), *140*
Pol, E. H., 3(21), *59*
Posner, G. H., 153(56), *183*
Pouchot, O., 85(85), *111*
Powell, R. E., 16(62), *60*
Powers, J. W. 1(6), 13(6), *59*
Prabhakar, S., 35(132), *62*
Priou, R., 93(100), *111*
Pritchard, N. K., 52(194), *63*
Pyun, H. Y., 165(85), *184*

R

Rabisohn, Y., 27(94), 48(94), 49(94), *61*
Raciszewski, Z., 114(10), 122(10), *138*
Radlick, P., 66(2), 68(12), 73(12), 99(2), 106(2), *109*
Rado, M., 116(32), 120(32), 121(32), 125(32), 130, 131(32), 133(32), 135, 137, *139*
Ranjon, A., 85(85), *111*
Rao, C. B. S., 37(145), *62*
Rao, V. S., 75(4)4, *110*
Ray, P. K., 29(102), 48(102), *61*
Rees, C. W., 17(63), *60*
Reese, C. B., 4(24), *59*
Reich, H. J., 76(46), *110*
Reid, S. T., 84(82), *111*
Renner, U., 45(158), *62*
Rice, K. C., 28(97), *61*
Richardson, A. C., 31(107), 49(107), *61*
Rigaudy, J., 93(103), *111*
Rio, G., 80(58), 81(62, 64), 85(85), *110, 111*
Rojahn, W., 38(148), 40(150), *62*, 103(20), 104(20), 105(123, 128), *112*
Rosati, R. L., 134(126), *62*
Roser, O. M., 35(131), *62*
Rosenthal, A., 32(114), 48(178), 49(178), 55(210), *61, 63, 64*
Rouchaud, J., 116(33a–33c), 120(33b), 122(33a–33c), 125(33a–33c), 130, *139*
Rowland, A. T., 1(12), *59*
Russell, G. A., 13(54), 20(54), 42(54), *60*
Rust, F. F., 144(36), *183*
Rus'yanova, N. D., 169(93), *184*

S

Sachdev, K., 36(140), *62*
Saeed, M. A., 1(11), *59*
Saeki, H., 26(86), *61*
Saito, I., 90(94, 95, 96), 107(96), 108(134), *111*, *112*
Saito, M., 31(108), *61*
Saito, Y., 44(156), *62*
Sakan, T., 109(138, 139), *112*
Sakata, T., 28(95, 96), *61*
Salo, W. L., 27(90), *61*
San Filippo, J., Jr., 153(54, 55), *183*
Santosusso, T. M., 2(14), 13(14), *59*
Sato, T., 44(156), *62*
Saxton, J. E., 51(191), *63*
Schade, G., 95(109, 110), 103(110), 104 (110), 105(129), *112*
Scheffer, J. R., 69(25), *110*
Scheit, K. H., 2(16), 13(16), *59*
Schenk, G. O., 67(3), 69(13, 15), 71(13, 35), 77(47, 48), 78(49–52), 79(53), 80(57), 81(13), 82(47, 67, 68, 69), 84(74, 76–80), 85(77), 91(97), 93(13), 94(13), 95(107, 109), 100(13, 49), 101(79), 102(13, 15, 79), 103(79, 119, 121), 104(119), 108(135), *110*, *111*, *112*
Schlessinger, R. H., 91(99), 92(99), *111*
Schloz, V., 4(26), *59*
Schneider, F., 23(72), *60*
Schneider, R. A., 35(136), *62*
Schneider, R. S., 37(147), *62*
Schmid, H., 45(157, 158), *62*
Schmidt, C., 165(86), *184*
Schmidt, R. R., 4(26), *59*
Schmidt, W., 84(83), *111*
Schmidt-Thomé, J., 23(70), *60*
Schofield, K., 4(24), *59*
Scholl, M. J., 85(85), *111*
Schrieber, J., 22(68), *60*
Schroeter, S. H., 82(67), 95(107, 109), *111*, *112*
Schuller, W. H., 100(113), *112*
Schulman, J., 83(72), *111*
Schulte-Elte, K. H., 67(3), 79(56), 94(3), 103(121), *109*, *110*, *112*
Schwille, D., 4(26), *59*
Scoffone, E., 40(151), *62*
Scott, A. I., 105(126), *112*
Sebera, D. K., 145(37), *183*
Sehon, A. H., 165(86), *184*
Sergeev, P. G., 114(9), 122(9), *138*
Seshardri, T. R., 51(192), *63*
Setaka, M., 145(40), *183*
Shapiro, R., 4(24), *59*
Sharp, D. B., 68(5), *109*
Shavel, J., 35(133), *62*
Shechter, H., 87(88), *111*
Sheehan, J. C., 10(42, 43), *59*
Shelton, J. R., 172(97), *184*
Sheng, N. M., 116(25, 30), 117(25, 35), 118(25), 120–127(25, 30, 35), 130(25, 30), 131(25), 132(30), 134(30), 135, 136, 137, *139*, 176(114), *185*
Shibata, H., 31(109), *61*
Shiro, M., 53(198), *63*
Shyrock, G. R., 27(91), *61*
Simonart, P. C., 27(90), *61*
Slessor, K. N., 48(173), *63*
Skingle, D. C., 50(189), *63*
Skold, C. N., 91(99), 92(99), *111*
Smejkal, J., 29(100), *61*
Smissman, E. E., 38(149), 50(149), *62*
Smith, H., 47(163), 50(163), *63*
Smith, S. G., 13(52), *60*
Sprake, M., 116(27), 117(27), 119(27), 120–123(27), 125(27), 130(27), 131 (27), 133(27), *139*
Sonnenberg, J., 88(91), *111*
Sonnet, P. E., 33(122), 34(122, 127), *61*, *62*
Šorm, F., 28(99), *61*
Sosnovsky, G., 142(15, 16, 17, 20), 147 (46), 163(17), 169(92), 171(15, 16, 17, 46), 172(15, 17, 95), 173(17), 175(100, 103), 176(16, 17, 105, 109), 178(105, 109), 179(105), 180(95, 105, 109), 181(92, 105), *182*, *183*, *184*
Sowa, W., 29(104), 48(171), 49(171), *61*, *63*
Stacey, M., 27(93), *61*
Stammer, C. H., 11(45), *59*
Stevens, C. L., 55(208), *64*
Stevens, H. C., 114(11), 122(11), *138*
Stewart, R., 143(27), 144(27), *182*
Stiller, K., 89(93), 91(93), 108(93), *111*

Story, P. R., 69(26), *110*, 155(57, 58, 59), 174(57, 58), *183*
Strehlow, W., 91(98), 92(98), *111*
Subramanian, R. V., 158(68), 161(68, 76), 171(76), *183*, *184*
Sussman, S., 113(16), 114(16), *138*
Svensson, S., 33(119), 50(119), *61*
Sweat, F. W., 1(5), 16(60), 21(60), 47(60), *59*, *60*
Swern, D., 2(14), 13(14), *59*, 126(40a, 40b), *140*
Syhora, K., 7(37), 21(37), *59*
Syz, M., 115(16), *139*
Szarek, W. A., 26(83), 50(83), *60*

T

Takashita, I., 31(109), *61*
Takatsu, A., 44(156), *62*
Tanida, H., 156(60, 61), 157(61), 174(60), 183
Tanner, D. W., 75(41), *110*
Tatchell, A. R., 29(102), 48(102), *61*
Taub, D., 21(67), *60*
Taube, H., 143(30), 144(30), 145(37, 38), *182*, *183*
Taylor, K. G., 55(208), *64*, 69(22), *110*
Temple, C., 37(146), *62*
Tener, G. M., 2(19), 3(21), 4(23), *59*
Theander, O., 24(75), 29(105), *60*, *61*
Theilacker, W., 84(83), *111*
Thomas, G. H. S., 29(104), *61*
Thomas, R., 53(199), *63*
Tien, J. M., 1(7), 2(7), 13(7), *59*
Tittensor, J. R., 4(25), *59*
Todd, Lord, 4(24), *59*
Todd, P. F., 158(70), *183*
Toennies, G., 164(79), *184*
Tohyama, T., 44(156), *62*
Tolley, M. S., 27(93), *61*
Tong, G. L., 48(172), *63*
Torsell, K., 14(58), 16(58, 61), 42(58), *60*
Trecker, D. J., 79(55), 98(111), 106(111), *110*, *112*, 142(25), 164(83), 171(25), 175(25), 176(25, 113), *182*, *184*, *185*
Treibs, W., 113(2), *138*, 172(96), *184*
Tronchet, J. M., 26(84, 85), *61*
Turnbull, P., 7(37), 21(37), *59*
Tsao, M. S., 143(31), *182*
Tsugi, T., 2(13), 13(13), *59*
Tsuji, T., 43(155), 156(60, 61), 157(61), 174(60), *183*
Tsuyino, T., 43(155), *62*

U

Uhde, G., 103(118), *112*
Uri, N., 141(1), *182*

V

VanDyke, M., 52(194), *63*
Van Sickle, 115(15, 16), *139*
Verheyden, J. P. H., 25(80), 29(100), *60*, *61*
von E. Doering, W., 5(32), 13(51), 21(32), 44(32), 56(32), 57(32), *59*, *60*
Von-Gustorf, E. Koerner, 108(135), *112*
von Philipsborn, W., 45(157), *62*
Vitols, E., 26(82), *60*
Vizsolyi, J. P., 4(23), *59*

W

Walker, R. T., 4(27), *59*
Wallbank, B. E., 35(129), *62*
Walling, C., 116(20), *139*, 147(44), 155(44), 160(44), 171(44), 174, *183*
Wang, P. Y., 33(118), 50(118), *61*
Wasserman, H. H., 69(25), 83(71, 73, 77), 85(77), 86(86, 87), 87(87), 88(89), 89(93), 91(93, 98), 92(98), 94(104), 108(93), *110*, *111*, *112*
Waterhouse, D. F., 35(129), *62*
Waters, W. A., 71(32), *110*, 141(2), 143(26), 144(33), 168(91) 175(26), *182*, *184*
Weaver, W. M., 1(6), 13(6), *59*
Webb, R. F., 113(5), 122(5), *138*
Wehrli, H., 50(188), *63*
Weimann, G., 11(47), *59*
Weinges, K., 53(197), *63*
Weise, A., 1(1), *58*
Weinshenker, N. M., 36(141), *62*
Weisner, K., 35(138), *62*
Weiss, J., 141(3, 4), *182*
Welsch, J. E., 164(78), *184*
Wendler, N. L., 21(67), *60*

Wesseley, F., 2(17), 53(204), *59*, *64*
Westheimer, F. H., 22(69), *60*
Wexler, S., 68(6, 7), 70(6), 84(74), *109*, *111*
White, D. M., 88(91), *111*
White, E. H., 88(92), *111*
Whitesides, G. M., 153(54, 55), *183*
Whitham, G. H., 148(49), 173(49), *183*
Whitlock, R. F., 68(9), *109*
Wigfield, D. C., 9(38), 23(38), *59*
Wightman, R. H., 2(15), 13(15), *59*
Wikholm, R. J., 53(196), *63*
Wilberg, K. B., 143(28), *182*
Wild, J., 37(147), *62*
Willhalm, B., 79(56), *110*
Williams, A. H., 75(42), *110*
Williams, P. H., 114(8a), 122(8a), *138*
Williamson, A. R., 4(27, 28), *59*
Willmund, W. D., 79(54), *110*
Wilson, D. A., 46(160), 47(160), *62*
Windaus, A., 65(1), *109*
Winstein, S., 13(52), *60*
Wirtz, R., 78(52), *110*
Wise, L. D., 36(139), *62*
Witzke, H., 68(10), *109*
Wolf, P. F., 129(39), 137(39), *140*
Wolfrom, M. L., 33(118), 50(118), *61*
Wood, G. W., 81(63), *111*
Woods, L. A., 153(53), *183*
Woodward, R. B., 11(46), *59*
Wright, A. C., 68(9), *109*
Wuesthoff, M. T., 84(74), *111*

Y

Yamamoto, S., 45(159), *62*
Yang, N. C., 115(12), *138*, 142(17), 163(17), 169(92), 171(17), 172(17), 173(17), 176(17, 92), 181(92), *182*, *184*
Young, D. P., 113(3), 122(3), *138*
Young, D. W., 105(126), *112*
Yuen, G. U., 32(112), *61*

Z

Zajacek, J. G., 116(25, 30), 117(25, 30), 120–127(25, 30, 35), 130(25, 30), 131(25), 132(30), 134(30), 135, 136, 137, *139*, 176(114), *185*
Zavitsas, A. A., 147(44), 155(44), 160(44), 171(44), 174, *183*
Zenerosa, C. V., 4(29), 11(29), 18(64a, 64b), *59*, *60*
Zetzsche, F., 9(39), *59*
Ziegler, F. E., 33(122), 34(122, 127), *61*, *62*
Ziegler, K., 78(51), *110*
Zimmerman, H. K., 27(91), *61*
Zinnes, H., 35(133), *62*
Zissis, E., 49(180, 183), *63*
Zu Reckendorf, W. M., 48(167), 55(207, 209), *63*, *64*

SUBJECT INDEX

A

Acetal functions, effect on oxidation of adjacent groups, 28
Acetic anhydride, and DMSO, 19, 20, 22, 24, 29, 31–35, 38, 40, 42–54
Acetone, as solvent in epoxidation reactions, 129, 130
3′-*O*-Acetylthymidine, and DMSO–DCC mixtures, 4
Activation complex, bridged, 143, 145
Acyloins, oxidation of, 52
N-Acylureas, as byproducts, 7, 8
Alcohols, as solvents for epoxidation reactions, 128
Aldehydes, synthesis from primary alcohols, 23
Aliphatic hydrocarbons, as solvents for epoxidation reactions, 128
Alkaloids, oxidation of,
  with DMSO–acetic anhydride mixture 45
  with DMSO–DCC mixture, 33
Alkoxysulfonium intermediates, 2, 16, 42, 46, 54
Alkyl hydroperoxides, as oxidants in metal-ion-catalyzed reactions, 159, 160, 161
Allylic alcohols,
  epoxidation of, 137, 138
  oxidation of, 39, 56, 57
Amides, metal-ion-catalyzed oxidation of, 171
Androst-4-ene-3, 17-dione, from testosterone, 5–9
Androst-5-ene-3, 17-dione, from 3β-hydroxyandrost-5-en-17-one, 9
Androst-4-en-19-ol-3, 17-dione, DMSO–DCC oxidation of, 23
Androst-5-en-3β-ol-17-one, DMSO–DCC oxidation of, 23
Aromatic hydrocarbons, as solvents for epoxidation reactions, 128
Ascaridole, from α-terpinene, 78

B

Boric acid in epoxidation reactions, 122
α-Bromoesters, and DMSO, 1
sec-Butylhydroperoxide, as a source of singlet oxygen, 70
t-Butylhydroperoxide,
  availability, 177
  in oxidation of hydrocarbons, 166, 170
  in oxidation of nitriles, 176
  in oxidation of olefins, 171
  in oxidation of sulfides, 175
t-Butylperacetate,
  availability, 177
  as oxidant, with metal-ion catalyst, 171
t-Butylperbenzoate,
  in oxidation of sulfides, 176
  availability, 177

C

Cantharidin, synthesis by photooxidation, 78
Carbodiimides, care in use, 10
Carbohydrates, oxidation of,
  with DMSO–acetic anhydride mixture, 47
  with DMSO–DCC mixture, 24
Carbon disulfide as a solvent in photo-oxygenation reactions, 75, 76

Carbon-monoxide effect on epoxidation reactions, 136
Carbonium ion formation, in metal-catalyzed peroxide oxidations, 149–151, 153, 155, 157
Δ^2-Carenes, and photooxygenation, 95
Chloresterol-7α-d, photooxygenation of, 95
Chloresterol-7β-d, photooxgenation of, 95
Chloroformates, and DMSO, 2, 13
Cholane-24-ol, DMSO–DCC oxidation of, 23
Cholestanol, oxidation of,
  with DMSO–acetic anhydride mixture, 47
  with DMSO–sulfur trioxide mixture, 56
Cholesterol, DMSO–DCC oxidation of, 23
Chromic oxide, in pyridine, oxidation with, 28, 29, 35
Chromium in epoxidation reactions, 113, 131
Chromium trioxide, as catalyst, 113
Cobalt, in epoxidation reactions, 120, 131, 134
Cobalt, traces, violent reaction with peracids, 164
Cobalt (II), in decomposition of peracids, 165
Cobalt (III),
  in atom-transfer reactions, 144
  in electron transfer without atom transfer, 145
Copper,
  in decomposition of diacylperoxides, 142
  in epoxidation reactions, 131
Copper (I),
  complexes with olefins, 158
  in decomposition of organic peroxides, 146, 147, 153
Copper (II),
  in atom-transfer reactions, 144
  in decomposition of organic peroxides, 146, 151, 153, 155
  in oxidation of cycloheptane, 166
Copper (II), *cont.*
  in oxidation of hydroperoxides, 147
Cumene hydroperoxide, availability, 177
Cyclic peroxides from photooxidation reactions, 79
Cyclohexanone regeneration, as a source of singlet oxygen, 69
Cyclohexene, autoxidation of, 133
Cyclopentadiene, photooxidation of, 79

D

1-Deuteriobutyraldehyde, from 1,1-dideuteriobutanol, 15
Diacyl peroxides, as oxidants in metal ion-catalyzed reactions, 161–163
Dialkyl peroxides, as oxidants in metal ion-catalyzed reactions, 164
Diazonium salts, and DMSO, 2, 13
1,1-Dideuteriobutanol, DMSO–DCC oxidation of, 15
Diels–Alder reaction, 66, 83, 93, 98
Dihydrobenzanthracene diol, oxidation of, 52
Dimethyl sulfoxide (DMSO), epoxidation of, 122
Diphenyl sulfoxide, and acetic anhydride, 44
9,10-Diphenylanthracene-9,10-peroxide, as a source of singlet oxygen, 69
1,4-Diphenylcyclopentadiene, direct photooxygenation of, 80
Disulfides from thiols, 41
2′,5′-Di-*O*-trityl-3′-ketouridine, DMSO–DCC oxidation of, 32
3′,5′-Di-*O*-trityl-2′-ketouridine, DMSO–DCC oxidation of, 32
2′,5′-Di-*O*-trityluridine, DMSO–DCC oxidation of, 32
3′,5′-di-*O*-trityluridine, DMSO–DCC oxidation of, 32
1,2-Dioxetanes, as intermediates in photo-oxidation, 99, 100
Dye sensitizers for photooxygenation reactions, 72, 73

E

Electron transfer, between metal ions and solutions, 143
Electronically excited states, as intermediates in photooxgenation, 68
Enamines, photooxidation of, 99, 100
Endoperoxides, from photooxidation reactions, 78, 80, 85, 101
Epimerization, in the presence of adjacent benzamido functions, 30, 31
Epoxides, and DMSO, 2, 13
Ergosterol, cyclic diene, photooxidation of, 100
2-Ethylidenebicyclo [2.2.1] hept-5-ene, photooxygenation of, 98

F

Fenton's reagent, 141
Flavonols, DMSO–acetic anhydride oxidation of, 51
Furans, photooxygenation of, 81–84

H

α-Haloketones, secondary, and DMSO, 1, 2
Homoallylic alcohols, DMSO–DCC oxidation of, 23
Homoannular dienes, photooxidation of, 100
Homophthalaldehyde, from indene, 99
Hydrocarbons, synthetic oxidations of, 166
Hydrogen peroxide, as a source of singlet oxygen, 70, 71
Hydroperoxides,
  oxidation in presence of Copper (II), 147
  from photooxidation reactions, 66, 85, 88, 90, 94, 98
Hydroquinones, in photooxygenation reactions, 77
3β-Hydroxyandrost-5-en-17-one, DMSO–DCC oxidation of, 9
Hydroxyl groups, DMSO–DCC oxidation of, 34, 40
3α-Hydroxy-5-β-pregn-16-en-20-one, DMSO–DCC oxidation of, 21
3β-Hydroxy-5-β-pregn-16-en-20-one, DMSO–DCC oxidation of, 21
11-Hydroxysteroids, DMSO–DCC oxidation of, 21
11-Hydroxysteroids, DMSO–acetic anhydride oxidation of, 45

I

Imidazoles, photooxidation of, 88–91
Imidazolinone formation, mechanism of, 90–91
Indene, photooxidation of, 99
Indole akaloids, DMSO–DCC oxidation of, 33, 42
"Inner sphere" transition state, 143, 145, 148, 149, 150, 153
Inositols, DMSO–acetic anhydride oxidation of, 49
Iron, in epoxidation reactions, 131, 134
Iron, traces, violent reaction with peracids, 164
Iron (III), in atom-transfer reactions, 144
Isopropyl alcohol, benzophenone-sensitized photooxygenation, 67
Isotetralin, photooxidation of, 100

K

Kornblum oxidation, 13

L

Lactams, metal-ion-catalyzed oxidation of, 171
Limonene, optically active, photooxygenation of, 94, 97

M

Manganese, in epoxidation reactions, 120, 131, 134

Manganese, traces, violent reaction with peracids, 164
Manganese (II),
in decomposition of peracids, 165
in oxidation of amides, 171
Manganese (III),
in oxidation of sulfides and sulfoxides, 175
in peroxide oxidations, 155
Metaborates, in epoxidation reactions, 123, 126
Metal–hydroperoxy radical complexes, formation of, 145
Metal ion–hydroperoxide complexes, in epoxidation reactions, 135, 137
Methallyl alcohol, epoxidation of, 132
Methyl 4,6-benzylidene-2-deoxy-α-D-allopyranoside, DMSO–acetic anhydride oxidation of, 49
Methyl methylenesulfonium ion,
in DMSO–acetic anhydride mixtures, 45
in thiomethoxymethyl ether formation, 19
Methyl 2,3,4,6-tetra-*O*-acetyl-D-gluconate, from 2,3,4, 6-tetra-*O*-acetyl-β-D-glucopyranose, 54
Methyl 2,3,4-tri-*O*-acetyl-D-glucopyranoside, DMSO–DCC oxidation of, 27
3-Methyl-1-(2,6,6-trimethylcyclohexen-1-yl)-1,3-butadiene, photooxidation of, 101
3-Methylenecyclobutanone, synthesis of, 36
Molybdenum, in epoxidation reactions, 117, 120, 121, 126, 128, 129, 130, 131, 132, 134, 135, 136, 137, 138
Monodeuteriodimethyl sulfide, from 1,1-dideuteriobutanol, 15
Monohydroxyquinones, oxidation of, 53

N

Neoabietic acid, photooxidation of, 100
Neopentenyl, 35
Nickel, in epoxidation reactions, 117
Nickel, traces, violent reaction with peracids, 164
Niobium, in epoxidation reactions, 117
Nucleoside 5′-aldehydes,
epimerization in attempted chromatography, 26
oxidation, with phosphorus pentoxide, 55
synthesis, 25
Nucleosides, oxidation of, 4, 32

O

1-Octene, epoxidation of, 132
Olefins, metal-catalyzed epoxidation of, 113, 114, 116, 117
Olefins, photooxidation of, table, 102–109
Olefins, metal-ion-catalyzed oxidation of, 170, 171
"Outer-sphere" transition state, 143–145, 148
1,4-Oxa-2-cyclohexane, from the endoperoxide, 78
Oxazoles, photooxidation of, 86–88
Oxygen, effect on hydroperoxide epoxidation of olefins, 130, 131
Oxygen function, allylic, introduction by photooxidation, 94
Ozone, as a source of singlet oxygen, 69

P

Peracids,
metal, in epoxidation reactions, 113, 114, 117, 136
organic, caution in use of, 164
organic, in metal-ion-catalyzed oxidation, 164–166
Peresters, in metal-ion-catalyzed reactions, 163, 164
Peroxide formation, in photooxidation, 66
α-Phellandrene, photooxygenation of, 78
Phenacyl halides, and DMSO, 1
Phosphorous pentoxide, and DMSO, 24, 29, 32, 44, 54–56

Photooxygenation, mechanism, 68
Polyporic acid, oxidation of, 52
Polysaccharides, DMSO–DCC oxidation of, 33
Propylene, epoxidation of, 132
Pseudoaxial hydrogen, in steroidal olefins, 94
Pseudoequatorial hydrogens, in steroidal olefins, 94
Pseudourea,
  in addition of DMSO to DCC, 12, 14, 15
  alkoxysulfonium derivative of, 13
Pulvinic acid dilactone, from polyporic acid, 52
Pummerer rearrangement, 9, 13, 20, 42
Pyridine, and DMSO, mixture, 2
Pyridinium salts, as proton sources, 7, 33
Pyrroles, photooxidation of, 84–86

Q

Quinol acetates, and DMSO, 2

R

Rhenium, in epoxidation reactions, 117
Ruthenium tetroxide, in oxidation, 29

S

Secondary alcohols, oxidation, 36
Selenium, in epoxidation reactions, 117
α-Silylcarbinols oxidation of, 36, 37
Singlet oxygen, in photooxidation, 66, 69–71, 77, 88, 91–94, 98
Sodium borotritiide, reduction of sugars with, 49
Solvent, effect of, in metal-ion-catalyzed oxidation, 158–159
Solvents, for use in photooxygenation reactions, 74, 75
Steroids,
  DMSO–acetic anhydride oxidation of, 45
  DMSO–DCC oxidation of, 21
Substrate structure, in photooxygenation reactions, 76
Sulfides, metal-ion-catalyzed oxidation of, 175
Sulfoxides, metal-ion-catalyzed oxidation of, 175
Sulfurtrioxide–DMSO mixture, in oxidation reactions, 56–58

T

Temperature, effect on expoxidation reactions, 130
α-Terpenine, photooxygenation of, 78
Terpinolene, photooxidation of, 101
Tertiary alcohols, dehydration in presence of DMSO–DCC mixture, 38
Testosterone, oxidation of,
  with DMSO–acetic anhydride, 47
  with DMSO–DCC, 5–9, 11, 17, 21
  with DMSO–sulfur trioxide, 56
2,3,4,6-Tetra-*O*-acetyl-β-D-glucopyranose, oxidation of, 54
Tetracovalent sulfur, intermediate, 43
Tetramethylene sulfoxide, and acetic anhydride, 44
Thiols, reaction with DMSO–DCC, 41
Thiomethoxymethyl ethers, as oxidation byproducts,
  with DMSO–acetic anhydride, 45, 46, 48
  with DMSO–DCC, 7, 19, 25
  with DMSO–sulfur trioxide, 57
Thiophenes, photooxidation of, 91
Thiosulfonium intermediate, in DMSO–DCC oxidation, 41
Thymidine 5′-phosphate,
  derivatives, 3
  effect of DMSO on, 2, 3
  oxidation of, 40
Transannular proxides, from photooxidation reactions, 78, 79, 81, 83, 87, 90, 92, 93
Transition states, in metal-ion–solution oxidation-reduction sequence, 143

Trautz reaction, 71
Triplet oxygen, in photooxygenation, 67
Tungsten, in epoxidation reactions, 113, 117, 120, 122, 131
Tungstic acid, as catalyst, 113

V

Vanadium, in epoxidation reactions, 113, 116, 117, 120, 121, 128, 130, 134, 135, 137, 138
Vanadium traces, violent reaction with peracids, 164
Vanadium (III), in decomposition of peracids, 165
Vanadium pentoxide, as catalyst, 113, 116, 121
  in oxidation of aldehydes, 176
  in oxidation of sulfoxides, 175

Y

Ylid intermediates, 13, 15, 18, 19, 20, 42, 43, 54
Yohimbine, oxydation of,
  with DMSO–acetic anhydride, 45
  with DMSO–DCC, 34, 44

Z

Zirconium, in epoxidation reactions, 117